SpringerBriefs in Earth System Sciences

Series Editors

Gerrit Lohmann, Universität Bremen, Bremen, Germany

Justus Notholt, Institute of Environmental Physics, University of Bremen, Bremen, Germany

Jorge Rabassa, Labaratorio de Geomorfología y Cuaternar, CADIC-CONICET, Ushuaia, Tierra del Fuego, Argentina

Vikram Unnithan, Department of Earth and Space Sciences, Jacobs University Bremen, Bremen, Germany

SpringerBriefs in Earth System Sciences present concise summaries of cutting-edge research and practical applications. The series focuses on interdisciplinary research linking the lithosphere, atmosphere, biosphere, cryosphere, and hydrosphere building the system earth. It publishes peer-reviewed monographs under the editorial supervision of an international advisory board with the aim to publish 8 to 12 weeks after acceptance. Featuring compact volumes of 50 to 125 pages (approx. 20,000—70,000 words), the series covers a range of content from professional to academic such as:

- A timely reports of state-of-the art analytical techniques
- bridges between new research results
- snapshots of hot and/or emerging topics
- literature reviews
- in-depth case studies

Briefs are published as part of Springer's eBook collection, with millions of users worldwide. In addition, Briefs are available for individual print and electronic purchase. Briefs are characterized by fast, global electronic dissemination, standard publishing contracts, easy-to-use manuscript preparation and formatting guidelines, and expedited production schedules.

Both solicited and unsolicited manuscripts are considered for publication in this series.

Andrejs Zarins

The Geology of Agate Deposits

 Springer

Andrejs Zarins
Geological Society of America
Albuquerque, NM, USA

ISSN 2191-589X   ISSN 2191-5903 (electronic)
SpringerBriefs in Earth System Sciences
ISBN 978-3-031-67928-5   ISBN 978-3-031-67929-2 (eBook)
https://doi.org/10.1007/978-3-031-67929-2

This Springer imprint is published by the registered company Springer Nature Switzerland AG
The registered company address is: Gewerbestrasse 11, 6330 Cham, Switzerland

If disposing of this product, please recycle the paper.

To: Roger Pabian and John Boellstorff; and to my family, Barbara, Julian, Elizabeth, and Joanne O Connell

Preface

"Agate" is defined by the American Geological Institute Glossary (1960), "as a kind of silica consisting mainly of chalcedony in variegated bands or other patterns commonly occupying vugs in volcanic and some other rocks". Dana's Textbook of Mineralogy (1898) defines agate as a variegated chalcedony. Some definitions, terminology, associated mineralogy and petrology, geochemistry, and ore geology terminology are described, particularly opal and its variants; chalcedony; chert; zeolites and zeolitization, the role of metasomatism and thundereggs; geothermometric stable oxygen and hydrogen isotope and fluid inclusion studies; and cathodoluminescence (CL) microscopy.

There are three main types of agates: "volcanic" agate, hydrothermal vein agates, and sedimentary agates. Hydrothermal vein agates are found in breccia cracks and fissures of the host rocks, usually in volcanics. They form from a mineralized hydrothermal solution and are often related with ore deposits. Sedimentary-type agates form in relatively shallow sediments, such as lagoons or a regressive marine shelf environment in saline waters, at surface conditions with temperatures below 40 °C. Silica fills in cavities, vugs, and other empty pore spaces in sedimentary rocks such as limestones, siltstones, argillaceous carbonates, and claystones. Volcanic agate deposits occur in large igneous provinces in intraplate locations, where the agate deposits are usually associated with continental flood basalt provinces. They are chiefly found in tholeiitic, calc-alkaline igneous provinces, basalts, and andesites.

Agates are the products of intense hydrothermal metamorphism, alteration, and leaching of tuffs and ashes, at the surface and/or at near surface depths, involving alkaline waters at temperatures of 20 °C to 250 °C. The fluids involve near surface ate phase hydrothermal fluids, and mixed hydrothermal, vadose, and surface meteoric waters. Lebedev, 1967, described a silica metacolloid deposit in the andesitic-dacitic volcanics of the Pauzhetka Springs, southern Kamchatka Peninsula, Russia. This is the only known instance, in the world, of a silica gel deposit, that is a precursor to agate formation.

In many agate localities, however, the mineralogy, textures, and the broad areal extent of agates, with the absence of hydrothermal alteration indicators, indicate low temperatures of agate formation, with the alteration of volcanic glass involving low temperature mobilization, transport, and re-deposition of silica as H_4SiO_4 in the cavities of the host rock. Another source of silica is derived from the alteration of feldspars in arkosic arenites, feldspathic sandstones, and volcaniclastic sediments, the latter also including glassy rocks. Brazilian agates and opals in particular are thought to be derived from these source rocks. Agates are indicators of unconformities through the erosion, weathering, and alteration of the host and source rocks. The iron oxides found indicate pre- or syngenetic formation, and occur at the interface of host rock, in the agate, and accumulate in bands in the agate.

The multiple stage diagenesis of silica has been recognized as a series of complex dissolution-precipitation events. The diagenetic sequence is characterized by increasing crystallinity, crystal size, and structural order. The diagenesis of silica proceeds as silica in solution as H_4SiO_4, to a sol, then to a gel, then to an amorphous precipitate, to Opal A, to Opal-CT, to chalcedony, granular microcrystalline quartz, and finally to mega quartz. Banding seems to be due to the variations of the silica phases and their impurities. The most common impurity is red hematite. Minerals derived from weathering or alteration minerals, if present, react through the BZ reaction process, with the silica gel, producing plumes and bands. The BZ reaction is a chemical non-equilibrium reaction that produces non-linear chemical oscillation patterns and is an example of non-equilibrium thermodynamics; it is an incomplete or partial chemical reaction that is never completed. Agate color is due to variations in crystal size, microstructure, porosity, and water content. Colors are due to the impurities in the precipitate. Ductile deformation zones of agate bands are thought to be either escape or infiltration tubes for silicic fluids. Paragenetic minerals are found in volcanic agates as mineral inclusions in quartz, intergrowths with quartz, or pseudomorphs. Calcite, aragonite, hematite, limonite, goethite, pyrolusite, psilomelane, and occasionally, magnesite, are the corresponding paragenetic minerals found in volcanic agates. Secondary zeolites, along with calcite and the clay minerals, celadonite, smectite, illite, and chlorite, formed and deposited in vesicles and joints from the alteration of volcanic glasses and host rock pyroxenes, and plagioclase, in stages, from late phase hydrothermal magmatic fluids that involved meteoric waters in the later stages. The time frame for the formation of agate is thus initially in hundreds or thousands of years for the formation of the colloid and the initial BZ reactions; millions of years for subsequent crystallization and diagenesis of the agate.

Closely related to agate deposition by mineralogy and mode of formation are chert, thundereggs, and petrified wood. Opal deposits are also examined, but they have a closer affinity to agate as they are a stage in silica diagenesis.

There are two schools of thought on the formation of chert. The first theory believes that chert formed from biogenic sources in oceans, namely from silica secreting organisms such as foraminifera, radiolarians, and diatoms. They supposedly use all the soluble silica supplied by rivers and streams that empty into the oceans. It is argued that these organisms only use about 10% of the soluble silica. Thus, the formula: biogenic opal→opal-CT→chert (granular microcrystalline quartz). The second theory states that chert formed from silicates of clays and zeolites, that had interacted previously with clays and bicarbonates. Thus, the formula in inorganic precipitation: silica in solution→amorphous silica→opal-CT→chalcedony/ granular microcrystalline quartz→megaquartz.

Thundereggs are found exclusively in rhyolites and ignimbrites in the glassy component of the ignimbrite or rhyolite flow, in the spherulites or lithophysae. The most well-known thundereggs are found in the base of the late Oligocene to Early Miocene John Day Formation, south of Pony Butte, Oregon, in a rhyolitic ash flow. Most thundereggs consist of an outer shell of pale-brown aphanitic rock that is composed of volcanic glass shards, fine ash, and collapsed pumice lapilli; all of which are altered to radially oriented sheaves of fibrous cristobalite and alkalic feldspar. The core consists of white to bluish-gray chalcedony that contains concentric and planar bands and dendritic mineral growths. The lithophysae were formed not by exsolved gases, but by water under pressure in the pores of the perthite and cristobalite spherulites, turning to steam, and thus creating the star-like cavities. Fluid inclusion studies of thundereggs indicate temperatures of formation from 134 °C to 186 °C with the mobilization and accumulation of silica into the spherulites or lithophysae during a late volcanic or hydrothermal activity phase or soon after. With the interaction of heated meteoric waters and hydrothermal fluids that dissolved the overlying and surrounding volcanic ash and tuffs to clays and silica-rich solutions, chalcedony and agate infillings were solidified and hardened in the lithophysae cavities, forming the thundereggs.

Petrified wood is the replacement of wood tissues by silica by a sequential silica diagenesis process: colloidal silica → opal A → opal CT → chalcedony → macrocrystalline quartz. It is technically a quartz pseudomorph after wood. The most well-known petrified wood deposit is located in east central Arizona, where the petrified forest is found in the Late Triassic Chinle formation. Downed trees accumulated in river channels and were buried by sediments and volcanic ash. Groundwater dissolved the volcanic ash and became a silica gel that replaced the woody organic material.

Twenty-four agate provinces have been described according to plate tectonics, stratigraphy, by the "agatization" process, secondary mineralization, hydrothermal activity, "agate" stratigraphy, and agate types. The physical and chemical characteristics of agate deposits are described and correlated with environments and conditions of deposition. Related silica deposits, including opal, chert, petrified wood, geodes, and thundereggs, are described as well.

Albuquerque, USA   Andrejs Zarins

Contents

List of Figures

Chapter 1
Introduction

"Agate" is defined by the American Geological Institute Glossary (1960) "as a kind of silica consisting mainly of chalcedony in variegated bands or other patterns commonly occupying vugs in volcanic and some other rocks." Dana's Textbook of Mineralogy (1898) defines agate as "a variegated chalcedony. The colors are either (a) banded; or (b) irregularly clouded; or (c) due to visible impurities as in moss agate, which has brown moss-like or dendritic forms, as of manganese oxide, distributed through the mass. The bands are delicate parallel lines, of white, pale, and dark brown, bluish, and other shades; they are sometimes straight, more often waving, or zigzag, and occasionally concentric circular. The bands are the edges of layers of deposition, the agates having been formed by a deposit of silica from solutions intermittently supplied, in irregular cavities in rocks, and deriving their concentric waving courses from the irregularities of the walls of the cavity. The layers differ in porosity…" Agate is considered a semi-precious gemstone, prized by collectors.

I have tried to describe each agate province according to plate tectonics, the stratigraphy, the "agatization" process, secondary mineralization, hydrothermal activity, "agate" stratigraphy, and agate types. The physical and chemical characteristics of agate deposits are described and correlated with environments and conditions of deposition. I have also included related silica deposits as well, including opal, chert, petrified wood, geodes, and thundereggs. I have selected examples and illustrations, geologic stratigraphic columns and maps, and photographs of petrographic thin sections in order to emphasize structural, chemical, and temporal controls. Accordingly, I have included plate tectonic–lithotectonic relationships in order to better explain and interpret the volcanic processes involved in agate deposits.

The science of agates does not exist in a vacuum. I have included brief descriptions of earlier work on agates, those of Noggerath (1850), Liesegang (1915), Ostwald (1896), and Bauer (1904), upon which the foundations of agate theory are based.

Hydrothermal water, frequently mentioned in this book, is simply that, "hot water," regardless of origins. I have included definitions and explanations of

A. Zarins, *The Geology of Agate Deposits*,
SpringerBriefs in Earth System Sciences, https://doi.org/10.1007/978-3-031-67929-2_1

geothermometric (fluid inclusion) studies, isotope studies, and cathodoluminescence investigations and their relationships to agate genesis.

The geology of some of the agate provinces is conjectural, that is, the location is unknown or kept secretive and the geology can only be inferred. This is especially true of the Argentinian Condor agate; the Botswana agate is somewhat secretive. Not all the agate provinces have been discovered yet. New locations are probably yet to be found in the Russian Siberian flood basalt region; in West Texas, extending westward from the Cathedral Mountain area, and in the Chilean-Argentine Andes Mountains' and Andes Mountains' front range, among others. The Trans-Caucasus region agate bears looking into. It is a historical agate-producing area dating back to Roman times, in which little or no data is available.

Chapter 2
Timeline and History of Agates

The history of agates starts out with the earliest of civilizations, the Sumer civilization of roughly 3000–2300 B.C. Although Sumerian necklaces, cylinder seals, and ceremonial bowls and axe-heads of agate are found, it appears that the main use of agate was in the manufacture of cylinder seals. The source of the agate is probably from two adjacent areas, namely, Iran and Yemen. The later Babylonians, c. 1500 BCE, also used agate as tools and as bowls, cups, cylinder seals, amulets, and ornamentation (Fig. 2.1). In Europe, in a significant discovery in 2017, a Minoan agate seal stone of the Mycenaean era from the Aegean Bronze Age was found at the Griffin Warrior tomb near the Palace of Nestor, Pylos, Greece (Fig. 2.2). This is the Pylos Combat agate and is the best example of glyptic art from the Aegean Bronze age period. It is minutely and intricately carved.

In ancient Greece, Theophrastus, c. 371–287 BCE, a Greek philosopher and successor to Aristotle in the Peripatetic school, referred to agates in his treatise, "On Stones" as derived from the Achates River (the modern-day Dirillo River) in Sicily; hence the etymology of the word, "agate" (Greek, αχατης).

Examining the geology of Sicily, one sees that the geology of Sicily fits into the archetypical model of volcanic agate formation and occurrence (Fig. 2.3). The river Dirillo (or the river Achates) runs 54 km from its origin as springs from the Hyblaen Plateau to the entrance of the Sicilian Channel, southeast of Gela. "Dirillo" comes from the Arabic "Wadi Ikrila," the "River of Acrille," a nearby ancient Greco-Roman town; the Arabs conquering and settling Sicily in the Middle Ages, 827 A.D. wresting control of Sicily from the Byzantines. In the late Miocene, thrusting occurred in the Sicilian orogen and in Calabria, where an accreting wedge formed along with associated extension- and back-arc magmatism. This magmatism, occurring in the late Miocene to early Pleistocene, formed alkali-tholeiitic basalts from fissure eruptions and formed Mt. Etna, a stratovolcano composed mainly of basalts. Subsequent erosion of the agates from amygdaloidal basalts deposited agates in the gravels of the river.

A. Zarins, *The Geology of Agate Deposits*,
SpringerBriefs in Earth System Sciences, https://doi.org/10.1007/978-3-031-67929-2_2

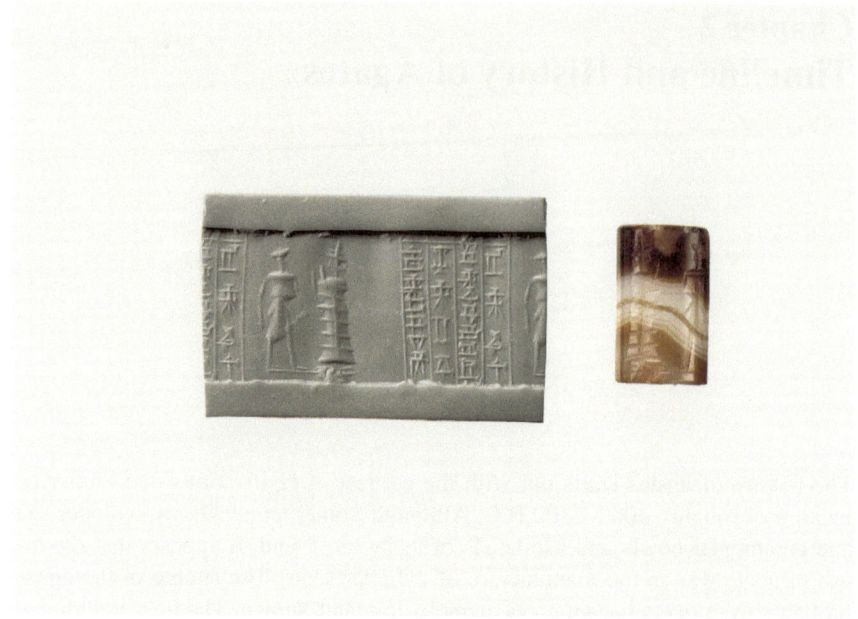

Fig. 2.1 Babylonian agate cylinder seal, first half, 2nd millennium BCE, Mesopotamia (public domain)

Fig. 2.2 Pylos Combat Agate Seal Stone, c. 1450 BCE, Late Minoan of the Mycenaean era, likely made in Late Minoan Crete; from the Griffin Warrior tomb near the Palace of Nestor, Pylos, Greece, 3.6 cm (Department of Classics, University of Cincinnati)

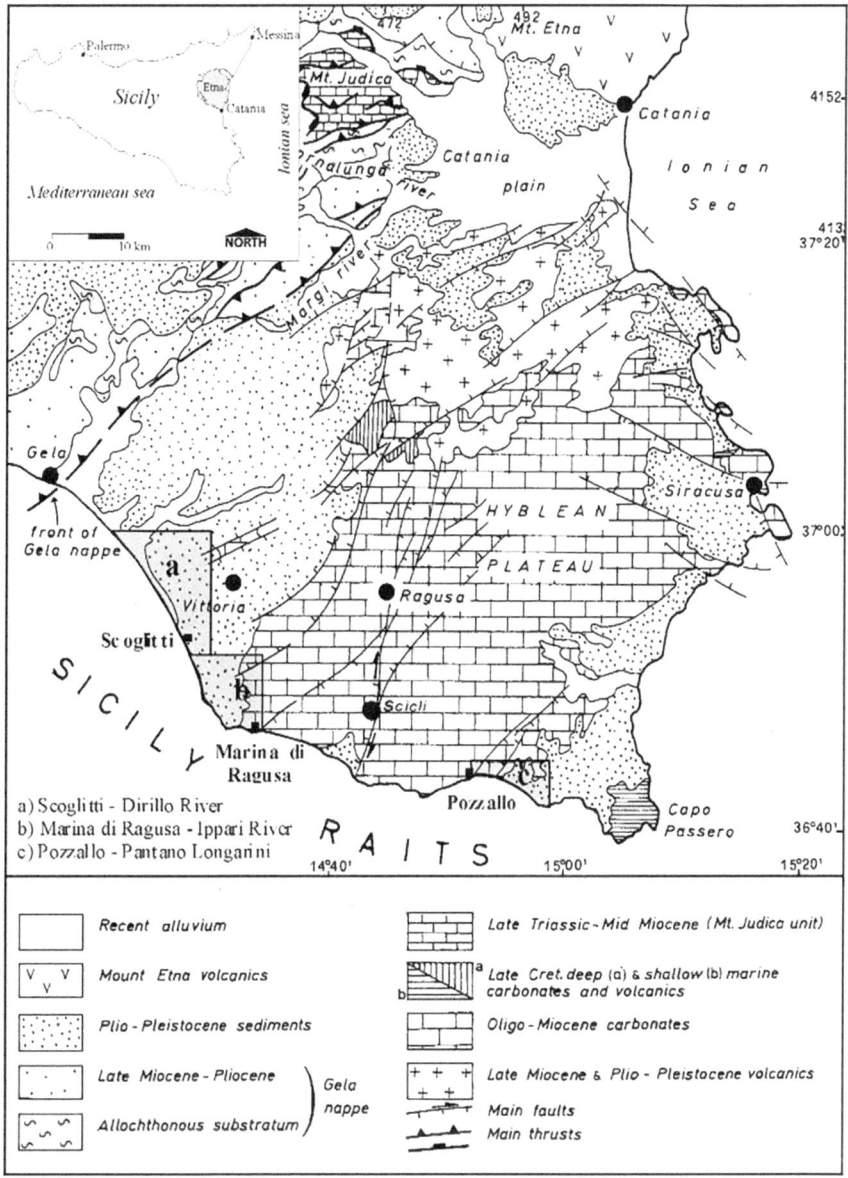

Fig. 2.3 Geologic map of Southeastern Sicily (after Ferrara and Pappalardo, 2004)

Pliny, 23/24-79 AD, Roman author, naturalist, and philosopher, made the same connection in his "Naturalis Historia," where he discusses agate localities (Crete, India, Phrygia, Egypt, Cyprus Mount Parnassus, Oeta Mountains, Lesbos, Messenia, Rhodus, and Persia), agate varieties ("jasper-like," "emerald agate," "blood agate," "white agate," "coralline agate," "tree agate," and "sacred agate"), agate superstitions (stopping storms, protection against scorpions, turn water cold, and invincible athletes), and artificial coloring of agates with honey. Pliny the Elder died in the Mt. Vesuvius eruption. The Romans cut agate into gems, beads, rings, cameos, ointment bottles, bowls, and cups. A two-handled agate cup from the time of Nero was presented to Charles the Bald of France in the ninth century.

In the Middle Ages, there are numerous references to agates worldwide. C. 650 A.D., Arab agate finger and signet rings, and talisman agates are described as "yamani" agate from Yemen and "Mahgreb" agate from Northwest Africa. C. 700 A.D., agate is found in Kan-su Province and in the mountains of northern Shan-si, Chi-li, and Shan-tung, China (and again in 1632, agate is discovered in Yiin-an and in the Ai-lao Mountains of Yung-ch'ang Province, China). In 1204 A.D. in Byzantine Constantinople, an agate bowl, 28 × 2 inches, created during the reign of Constantine I and thought to be the holy grail, was looted by Crusaders, and brought to Europe. In 1193–1280 A.D., Albertus Magnus, theologian, alchemist, and bishop of Cologne, Germany describes "agathes" in alchemist terms. Agates were thought to relieve snake and scorpion bites, and symbolize good health. In 1497, In Idar-Oberstein, Germany, an agate industry is formed and in 1500 A.D., Duarte Barbosa, a Portuguese explorer, found an agate industry in Limodra; Cambay; and Guzerat, India.

With the beginnings of science, especially geology, being formulated in the early Industrial period in Great Britain, France, and Germany, principally, various theories on the formation and occurrence of agates were formulated. In 1849, Wilhelm von Haidinger hypothesizes that agates form from solutions of silica that enter the rocks through diffusion. In 1850, Jakob Noggerath theorizes on the entrance/exit canals (tubes of entry or escape) in agates. In 1896, Wilhelm Ostwald formulates "Ostwald ripening": agates form in liquid sols, with smaller particles or crystals dissolving and redepositing on surfaces of larger crystals. He further describes flocculation as undissolved colloids that come out of suspension with the addition of a clarifying agent and forming a series of bands through the shrinking and drying of silica gel; frequently, there is a hollow in the interior with crystals of quartz. He further theorizes the quartz crystals cannot form in colloids on account of surface tension; when the tension is relieved in the hollow of the agate interior, complete crystallization can take place. In 1901, M. Forster Heddle, Professor of Chemistry at the University of St. Andrew, Scotland, describes characteristic features of agates (tube of escape, dilatation on the tube of escape, rent, agate dyke) and theorizes on agate formation. Max Bauer in 1904 theorizes that the tubes of escape and entry are formed by hot spring water that is saturated with silica, and intermittently saturates amygdaloidal rock forming the layers, bands, and tubes of entry or escape in agates. In 1915, Raphael Liesegang explains agate banding by experimenting with potassium dichromate in an agar gel and silver nitrate, whereby rings of silver dichromate are formed. He also discusses moss and dendritic agates, layered agates, and faulted agates.

Entering the modern period, agates enter the world of film and television, with the 1967 episode of "Star Trek" created by Gene Rodenberry (and a collector of agates), creating a Star Trek episode that features a silicon-based life form, the "horta," suggesting the possibility of silicon-based life forms in other worlds. In academia and popular rock-hounding literature, we have the following: in 1978, Roger Pabian describes various structures and inclusions in agates; in 1984, Landmesser publishes electron microscope images of agate bands and publishes a history and evaluation of past theories of agatization in "Das Problem der Achatgenese"; in 1989, MacPherson describes agates from the British Isles and suggests that banding forms when silica gel is separated into hydrous and anhydrous layers; in 1986, Wolter theorizes that agates formed by an infilled-cavity-by-percolation into an infilled cavity; and in 1989, Shaub presents the "syngenetic agglutinated colloidal silica" theory to replace the 200 year-old "epigenetic infilled cavity" theory.

Chapter 3
Background, Definitions, and Terminology

Before we examine and discuss the geology of various agate provinces, some definitions, terminology, associated mineralogy and petrology, geochemistry, and ore geology terminology are described in order that the reader may be able to understand the geology of agates. Since we are dealing with quartz and chalcedony, we are then also dealing with quartz that has, in addition to quartz itself, 11 crystalline and 2 non-crystalline polymorphs of SiO_2. Polymorphs by definition are minerals that are the same chemically, but differ in their molecular structure; that is, the manner in which atoms are arranged in the crystal structure. All polymorphs eventually convert to quartz. Only the polymorphs that occur frequently in our discussion of agates will be described.

3.1 α- and β-Quartz

α-quartz is a low-temperature, hexagonal, trapezohedral–tetartohedral polymorph that forms at temperatures below 573 °C. It occurs primarily in veins, geodes, and large pegmatites, while β-Quartz is hexagonal, trapezohedral–hemihedral and forms at temperatures from 573 to 870 °C. It forms in graphic granite, granite pegmatites, and porphyries.

3.2 α- and β-Tridymite

The α-variant of tridymite is a low-temperature form of many different structural and crystal classes, but is chiefly orthorhombic. It is a constituent of opal-CT. The high temperature, or β-variant, is a hexagonal form that forms above 800 °C in acidic volcanic rocks.

A. Zarins, *The Geology of Agate Deposits*,
SpringerBriefs in Earth System Sciences, https://doi.org/10.1007/978-3-031-67929-2_3

3.3 α- and β-Cristobalite

α-Cristobalite occurs as a transitional silica polymorph at low temperatures that is formed during the diagenesis of siliceous ooze sediments as tetragonal or as pseudo-isometric crystals. The high-temperature β-variant is isometric with octahedral crystals, which are often twinned or occur in crystalline aggregates, and are found in silica-rich volcanic rocks.

3.4   Opal

Opal is commonly associated with agates and is also the precursor to agate or chalcedony formation. Opal consists of highly disordered hydrous silica polymorphs that contain varying amounts of amorphous silica and tridymite.

Opal-A is highly disordered and nearly amorphous (Fig. 3.1). Type-A opals are the common opal, with irregular packing of different-sized spheres, and the gem opal, with close packing of equal-sized spheres forming an ordered network, allowing light to diffract into an array of rainbow colors or "play of color." They are the opals formed at low temperatures in sedimentary rocks that are amorphous, with 4–9 wt% H_2O, and are 1–8 μm in size. Opal-A is rare to absent in agates, with the non-opalescent variety found in hot springs.

Opal-CT is volcanic opal that forms at slightly higher temperatures than opal-A. It consists of non-crystalline spheres of cristobalite, 1–10 μ in size, that are mixed with crystalline, platy, blade-shaped crystals of tridymite (Fig. 3.2). Opal-CT is length-slow optically. It is the higher temperature form of opal found in volcanic rocks, where it is found filling the seams, fissures, and cavities of igneous rocks where it is deposited during the last rock cooling stages (Fig. 3.3). It contains 9–18 wt% H_2O.

Opal is also formed from the weathering and alteration of rhyolitic volcanic rocks: ignimbrites, volcanic ash, and tuffs. Most prevalently, opal forms from the siliceous skeletons of various sea organisms such as diatoms, radiolarians, and sponge spicules. Opal is also found as a precipitate, where in association with other evaporitic minerals such as dolomite, magnesite, and calcite/aragonite, it is precipitated out from high pH waters. Lowering of the pH causes the silica to precipitate out as opal.

3.5   Microcrystalline Quartz-Quartzine

Quartzine, as differentiated from chalcedony, consists of fibers, 100–200 nm in length, elongated parallel to the crystallographic c-axis, optically creating a high refractive index in this direction and is termed length-slow. Folk and Pittman (1971) suggest that quartzine is the result of deposition in sulfate-rich solutions in which

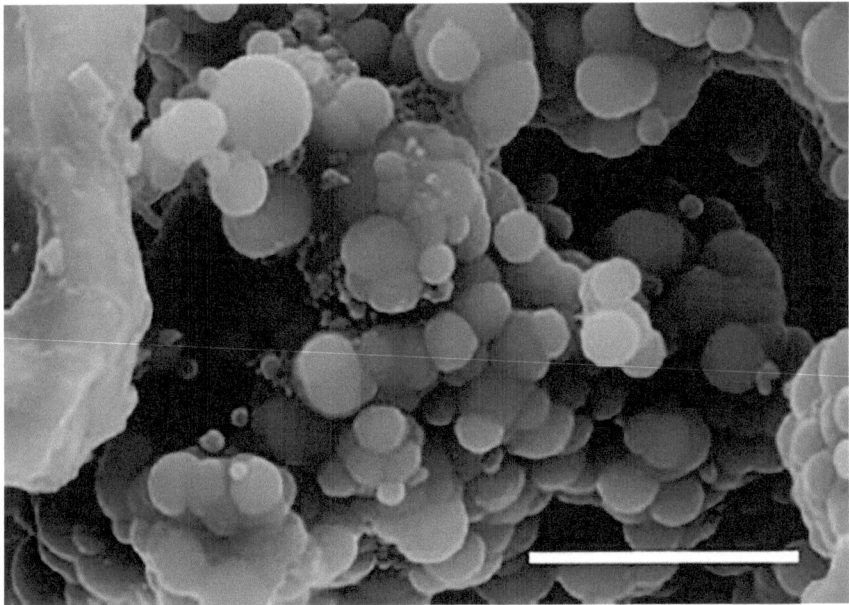

Fig. 3.1 SEM photograph of opal-A exhibiting typical 4 ± 8 µm diameter spheres; scale bar 10 µm (Lee, 2007)

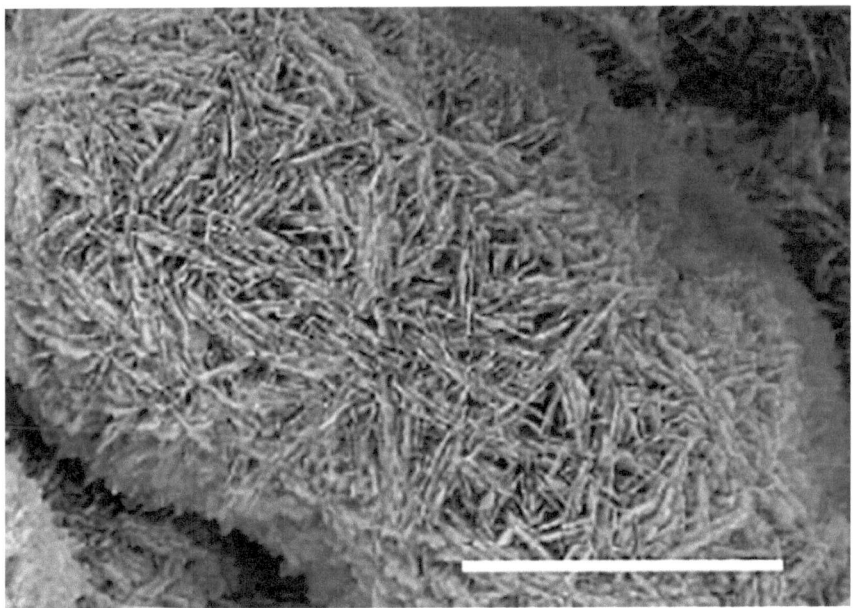

Fig. 3.2 SEM photograph of opal-CT showing thin-bladed crystals of tridymite, <10 µm in diameter; scale bar 10 µm (Lee, 2007)

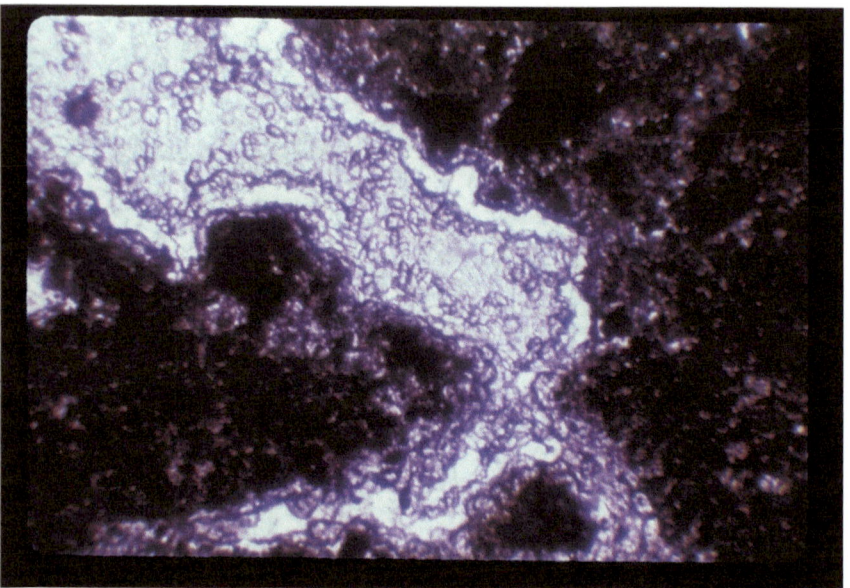

Fig. 3.3 Opal with microcollomorphic texture as vein filling in basalt; ×23, plane polarized light

silica sulfate ions prevent the silica from polymerizing into chains as in the common length-fast variety of chalcedony.

3.6 Moganite

Moganite is a monoclinic polymorph of quartz, moganite is found as intergrowths of fine-grained gray fibers with chalcedony in agate and chert. It consists of the alternate stacking of right- and left-handed quartz with periodic Brazil twinning. Pure moganite is rare. Chert, especially evaporitic chert, has the most moganite, up to 30–50%. Flint has 13–17% moganite concentration, agate has 5–20%, the siliceous corals in Tampa Bay, Florida average 20%, jaspers from the Iron formations have 0%, the weathered outer rinds of agates have 0%, and Arkansas Novaculite has 0%. Interestingly, agates older than 100–150 million years have no moganite. In addition, moganite formation and paragenesis seems to be associated with evaporite deposits (Heaney, 1995; Parthasarathy et al., 2001).

3.7 Microcrystalline Quartz-Chalcedony

Chalcedony consists of fibrous radiating bundles (Fig. 3.4). In general, the chalcedony fibers are an intergrowth of quartz microcrystals, 0.1–3 μm in size, which are orientated with the c-axis perpendicular to the fibers or length-fast direction. In length-slow quartzine, their orientations are parallel to the c-axis. Recent research by Wang and Merino (1999) shows fibrous chalcedony crystals that are polysynthetically twinned according to the Brazil twin law, with the c-axis twisted around the fiber axis, with the resultant arrangement of the atoms and molecules into spiral layers. The horizontally banded variety is composed of closely packed, radiating, spherulitic fibers with no evidence of c-axis twisting of fibers. The fibers are elongated in the direction of the a-axis, perpendicular to the c-axis as in quartz crystallites. This results in a lower refractive index in the direction of the fibers, which is a negative optical characteristic called length-fast. The fibers are typically ~100 to ~200 nm in size. This is the variety of spherulitic chalcedony that usually fills the vesicles and cavities of rock. Wall-lining chalcedony, on the other hand, is a variety that consists of parallel-fibrous aggregates in which the c-axis is twisted around the fiber axis, causing the distinctive, rhythmic banding of extinction. It is length-fast; >1 μm, typically 50–350 nm. Chalcedony cannot form above 180 °C as the microstructure of agate is destroyed under hydrothermal conditions above 200 °C (Graetsch, 1985).

Fig. 3.4 Spherulitic chalcedony in agate, ×23, polarized light

3.8   Fine Quartz

Fine quartz is a microcrystalline variety of quartz that has a granular texture and has a mutual orientation of strained grains, with sizes typically in the <5 to 20 μm range. Under crossed polars an undulatory extinction is displayed.

3.9   Crystalline Quartz (α-Quartz, Mega Quartz)

Under the polarizing microscope, α-quartz, or mega quartz, has uniform extinction. It has well-defined subhedral grain boundaries with prismatic crystals, is typically elongate parallel to the c-axis, and has a positive refractive index. It is typically 20–50 μm in size. It occurs as a recrystallization product of chert and chalcedony. It occurs as drusy quartz, authigenic quartz crystals, quartz overgrowths, and geode quartz. Metamorphosed chert is considered mega quartz.

3.10   Water

Water is found in opal and chalcedony as molecular H_2O and water in silanol groups as H_2OSiOH. Silanol water is mostly present in small amounts in Opal-A at 0.5–1.0 and 0.1–0.3% in microcrystalline opal; chalcedony has 0.2–0.9% silanol.

Both types are subdivided into two types. Type-A consists of isolated non-hydrogen-bonded molecules and hydroxyl groups trapped into the structure as fluid inclusions; typically, 1 μ in diameter. Type-B is strongly hydrogen-bonded water molecules or hydroxyls, either in the structure or on external and internal surfaces. In Opal-A, due to its porous nature, the total water present is 10–12%. It is present in fluid inclusions of the Type-A category. As silica polymorphs become more ordered the total water content decreases. Thus Opal-CT has 3–10% total water, Opal-C has 1–3% total water, and microcrystalline quartz (chalcedony) has 0.5–2.5% total water.

3.11   Devitrification

Devitrification is the process whereby glassy rocks (especially tuffs) form into definite minerals, usually to minute quartz (cristobalite, tridymite) and feldspar (orthoclase, sanidine, and sodic plagioclase). Vapor phase crystallization is a coarser devitrification crystallization process. It is the change of the glassy rock from a glassy state to a crystalline state after solidification, showing curved crystals and fluidal banding (Fig. 3.5). In deposits greater than 600 feet in thickness, granophyric crystallization occurs. The texture of the rock consists of irregular growths of blebs, patches, and

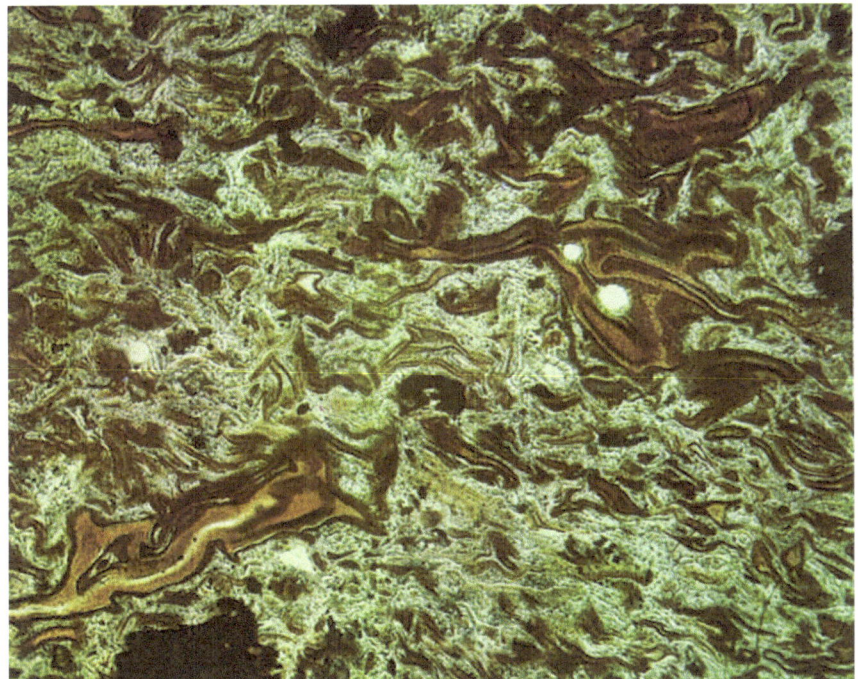

Fig. 3.5 Partially welded ash-flow tuff (Ross & Smith, 1961)

threads of quartz in a base of feldspar. Depending on the amount of water present, the thickness of the ash-flow tuff, and the temperature, devitrification is completed in thousands of years to millions of years. Marshal (1961) found that a volcanic glass to be converted to a felsite, with little or no water, devitrification took at least one million years at 300 °C; at 400 °C, a few thousand years. With water present, at low temperatures of around ~20 °C, devitrification proceeded at a rate of 4–5 μ per *~100 My.*

3.12 Secondary Mineralization

Rocks and minerals formed as a consequence of the alteration of pre-existing minerals. Secondary minerals may be formed in situ as pseudomorphs or paramorphs.

3.13   Deuteric Alteration

Deuteric alteration is the alteration produced during the last stages of magma formation and is the direct result or consequence of the consolidation of the magma. The late
differentiate minerals usually involved are chlorophaeite, chlorite, calcite, zeolites,
opal, chalcedony, and maybe some tridymite and cristobalite. Deuteric alteration is
differentiated from alterations made from secondary changes, due to a later period
of alteration.

3.14   Zeolitization

Amygdule agates are frequently found with igneous zeolites. Zeolites are hydrous
silicates of aluminum sodium and calcium; with an easy and reversible loss of water
by hydration. They are analogous to feldspars. Zeolites are formed from the alteration of volcanic rocks by late-stage hydrothermal fluids and are strictly a secondary
mineral. The chemical elements necessary for the secondary minerals were derived
from the alteration of the host rock, e.g., volcanic glass, olivine, clinopyroxene, and
plagioclase. $\delta^{18}O$ and $\delta^{13}C$ data indicated that the early phases of the mineralization
involved primarily magmatic waters; with the later phases involving increasingly
greater amounts of meteoric water (Ottens et al., 2019). The specific type of zeolite
is associated with a specific volcanic glass and temperature. For example, chabazite
forms from basaltic glass at 50–100 °C and heulandite forms from rhyolitic glass
at 150–225 °C. Zeolites, such as clinoptilolite, are also derived from the weathering of tuffs and volcanic ash. Zeolitization is the replacement or introduction
through fluids in open-space filling, of zeolites, particularly in rocks containing
calcic feldspars and feldspathoids. Zeolites are found commonly in association with
copper deposits. Open space filling through replacement of primary minerals, the
mineralogy, textures, and absence of alteration, seem to indicate low pressures and
temperatures; near-surface pressures and temperatures perhaps less than 100 °C.

3.15   Metasomatism and Thundereggs

Metasomatism is the simultaneous solution and deposition of minerals through small
openings that are usually sub-microscopic, mainly by hypogene (ascending) waters,
creating solutions by which a new mineral of partly or differing composition may
grow in the body of an old mineral or mineral aggregate. Potassium metasomatism
involves formation of K-feldspar, adularia, or sanidine (Holzhey, 2001), and sodium
metasomatism involves formation of plagioclase, sodic albitization, or albite. Lithophysae are manifestations of metasomatism. They are the first stage of potassic
metasomatism with the formation of spherulitic concentric growths of cristobalite

Fig. 3.6 Devitrification spherulite in volcanic glass showing radial structure; plane light (ufgrs. br, Feb. 2022)

needles and perthite shells (Figs. 3.6 and 3.7). They form only in silica-rich rhyo-lites and obsidian. Water trapped within the spheroidal fractions of the growth is converted to steam as the magma cools and forms the star-shaped or asteriated cavi-ties within the rhyolitic rock. Hydrogen metasomatism is the last stage of meta-somatism that involves silica mobilization from the silicified spherulites to form chalcedony, agate, and macro-quartz within the cavity. There is interaction with both meteoric and primary waters at this stage, suggesting low temperatures of formation (ca. 70–130 °C). Scalenohedral calcite, if present, is considered to be co-genetic with silica.

3.16 Glass Alteration

A principal source of silica in agates is considered to be from volcanic ash and from weakly indurated tuffs with attendant alteration and diagenesis (Fig. 3.8). Volcanic ash shards dissolve readily, especially in high pH and temperature conditions and environments. The high surface areas of volcanic ash shards give rise to high solubil-ities and rates of solution that lead to supersaturation and the precipitation of silica or

Fig. 3.7 Devitrification spherulite as previous photo, but polarized light; showing radial feldspar and quartz (ufgrs. br, Feb. 2022)

siliceous oozes. Alteration of volcanic glass involves low-temperature mobilization, transport, and re-deposition of silica as agate infillings in the cavities of the host rock.

The alteration process of volcanic ash and weakly indurated tuffs involves the formation of phyllosilicate minerals, principally the clays (kaolinite, illite, montmorillonite, or bentonite) and zeolite through the secondary alteration processes of hydration, weathering, and oxidation, with silica released as a free ion as H_4SiO_4. The argillic alteration of volcanic glass by low-temperature hydration is a chemical reaction with meteoric waters in epigenetic conditions. This involves mobilization and losses of Na and Ca and leaching of silica. This process is similar to sericitization and chloritization of rocks. The mobilization of the alkalis elements produces a high alkaline environment for silica solution and the formation of the clays.

Fig. 3.8 Bandelier ash-flow
tuff, New Mexico; 27 ×
46 mm, ×2, plane light

3.17 Geothermometric Stable Oxygen and Hydrogen Isotope and Fluid Inclusion Studies

The oxygen ratios are the result of cyclical variations in the abundance of oxygen with an atomic mass of 18 to the abundance of oxygen with an atomic mass of 16 in water. Evaporation and the cooling of moisture in air favors the lighter isotopic values, O^{16} and H^1. As water migrates toward the higher latitudes or poles and the higher elevations across continents, the water becomes depleted in the heavier ^{18}O and D. The ratio is linked to water temperature of ancient oceans and ancient climates. O^{16} requires less energy to vaporize, while ^{18}O tends to concentrate in liquid. Thus, more energy is required to vaporize $H_2{}^{18}O$. Higher temperatures equal less $H_2{}^{18}O$, producing negative numbers, while colder temperatures equal more $H_2{}^{18}O$, producing more positive numbers. An increase in $\delta^{18}O$ reflects declining temperatures. Thus, high $\delta^{18}O$ values reflect lower temperatures.

The isotope ratios are expressed in delta units (δ) as parts per thousand or 0/00 with the differences (%) compared relatively to an arbitrary standard known as Standard Mean Ocean Water (SMOW):

$$\delta\ 0/00 = \left[(R - R_{Standard})/R_{Standard} \right] \times 1000$$

where R and $R_{standard}$ are the isotope ratios, $^2H/^1H$ or $^{18}O/^{16}O$ of the sample and standard, respectively. A mass spectrometer measures $^{18}O/^{16}O$ ratios. A fractionation factor (α) for quartz-water is corrected relative to SMOW and the appropriate geologic epoch meteoric value is calculated:

$$\alpha = \frac{\frac{1+\delta^{18}O_{quartzSMOW}}{1000}}{\frac{1+\delta^{18}O_{waterSMOW}}{1000}}$$

Then:

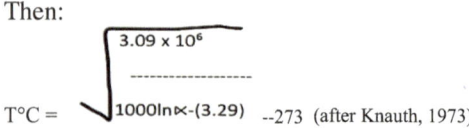

$T°C = \sqrt{\dfrac{3.09 \times 10^6}{1000\ln\alpha - (3.29)}}$ --273 (after Knauth, 1973)

The $\delta^{18}O$ and δD values are plotted into to summary diagrams which depict areas of the different types of waters (connate, meteoric, magmatic, hydrothermal, seawater) relative to SMOW and the kaolinite weathering line(s). Figure 3.9 depicts a typical summary diagram of the isotopic composition of waters ($\delta^{18}O$ 0/00 vs. δD 0/ 00) of an epithermal ore deposit in relation to the magmatic and metamorphic water boxes, meteoric water line, and SMOW (Zamanian et al., 2020). Temperatures are determined by the above-mentioned equation.

The oxygen isotope studies have shown that the water in agates exhibit strong variations in $\delta^{18}O$ values and suggest fluid mixing under non-equilibrium conditions with a non-crystalline precursor, i.e., a silica-rich solution or a gel and that exact calculations are not possible due to the unknown origins of the primary fluids.

Agate formation temperatures in Götze et al. (2020) range from 20 to 230 °C and include sedimentary-type agates. The paleotemperatures of Chihuahuan geodes and the Coyamito agate is 30–68 °C (Keller, 1977). Brazilian geodes and agates (Gilg

Fig. 3.9 $\delta^{18}O$ 0/00 versus δD 0/00 of an epithermal ore deposit (after Zamanian et al., 2020)

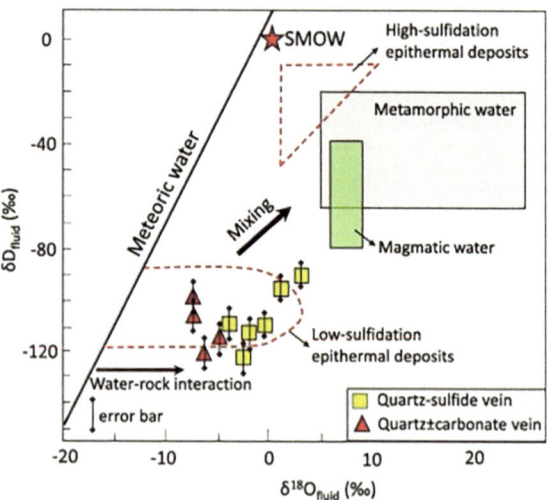

et al., 2003) are less than 100 °C, probably 50–80 °C. Etendeka volcanics agates (that are equivalent to the Brazil Parana basalt agates), from the Skeleton coast, Namibia, are 26–169 °C, averaging at 120 °C (Harris, 1989). The Scotland agates of Lower Devonian age, as calculated by Fallick et al. (1985), ~50 °C. Oxygen isotope $\delta^{18}O$ values of Southern Germany agates (macro-quartz and calcite), as calculated by Holzhey (2001), are 68–128 °C.

Water is present in agates as OH (Si–OH silanol groups). By heating a vacuole until a single phase (usually water in this case), the minimum temperature at the time of entrapment of the original fluid can be estimated. The water inclusions in agate temperature values indicate values between <100 and >500 °C. This more or less confirms the temperature formation values of the oxygen isotope studies.

With fluid inclusions, certain assumptions must be met: (1) the contents of the vacuoles were deposited from an original ore-bearing fluid; (2) there were no particular surface effects that caused an atypical fluid to be trapped; (3) there was no subsequent leakage either in or out; and (4) a heterogenous, trapping of inclusions must be assumed.

A caveat is applied to these isotope studies, viz., the memory of their original conditions may have been changed by post-depositional exposure to fluids that are isotopically different by secondary exchange of fluids from continued passage of fluids. Isotope measurements must be interpreted and applied with great care. Subsequent volcanism and cycles of eruption, and hydrothermal activity at many of the agate localities could have altered the original isotopic values, obscured the original low formation temperatures of agates, and increased the temperatures of formation.

3.18 Cathodoluminescence, Trace Element, and Rare Earth Element (REE) Studies

Cathodoluminescence (CL) microscopy is the analysis of light emitted by a mineral or gemstone when excited by an electron source. The concentrations of elements required to produce measurable peaks in a CL spectrum can be orders of magnitude lower than those measurable by X-ray microanalysis in energy-dispersive (EDX) or wavelength-dispersive (WDX) mode. Cathodoluminescence microscopy and spectroscopy can be used to reveal differences in the microstructure of macrocrystalline quartz and chalcedony in agates, enabling the visualization of zoning and other internal structures in agates and quartz. Based on the fact that the defects causing the different CL emissions reflect the variable physio-chemical conditions of formation, CL spectra of agates often differ from those of quartz from crystalline rocks. In the CL spectra of agates, at least three broad emission bands can be detected: a dominating red band at 650 nm, a yellow band at about 570 nm, and a blue band of mostly low intensity at about 450 nm.

CL spectroscopy can also be used for microanalysis by revealing the presence (or absence) of trace elements, with each trace element giving rise to luminescence at a different wavelength. Many different factors influence the intensity and wavelength of CL, including (1) the band-gap energy of the mineral; (2) the presence of structural defects (e.g., vacancies, dislocations); and (3) the presence and concentration of trace elements. These latter centers may introduce (mid-gap) energy states; alternatively, they may allow intra-ion energy transitions such as within the d orbitals of transition metals or the f orbitals of rare earth elements (REE). The electronic transitions between these mid-gap or intra-ion states can release photons whose wavelength corresponds directly to the change in energy. Thus, wavelength discrimination (in the form of CL spectroscopy) enables trace elements or defects to be differentiated. Meteoric waters tend to favor crystal elements such as the elements calcium and magnesium, and the ions, sulfate, and carbonate. Volcanic waters tend to favor the more fugitive elements such as boron and fluorine. Aluminum is the most frequent trace element in quartz and chalcedony (up to a few 1000 ppm).

The rare earth elements are a set of 17 metallic elements. These include the 15 lanthanides plus scandium and yttrium. REE is used to identify the composition of magmatic fluids and their partitioning and fractionization composition. Thereby, by composition identification of magmatic and hydrothermal fluids that formed the host rock of agates, the nature of the fluids that forms agates can be identified. REE thus far has revealed nothing of significance with REE in agates.

3.19 Agate Formation-Stratigraphic Setting-Host Rock Lithology-Modes of Occurrence

Agate formation has been a geological conundrum that Noggerath (1850), Ostwald (1896), Liesegang (1915), Ross (1941), and many others have attempted to solve. The basic theory is the rhythmic theory, based on the assumption that the bands in agate are similar to other rhythmically deposited phenomena, such as the classic lake sediment varves.

The hydrothermal or high-temperature rhythmic theory is the most commonly accepted theory for agate formation, viz., silica-rich hydrothermal fluids agatize vesicles and cavities within the host rock. Silica-rich fluids are injected into the cavity while the enclosing rock is still hot. Fluids are essentially a silica gel or sol which is rhythmically deposited into the vesicles by periodic infusion of the gel creating the bands or layers. Hydrothermal indicators are propylitization of the country rock and an appropriate hydrothermal mineral assemblage. This theory of agate formation suggests agates formed from super-critical fluids with water temperatures over 374 °C. With protracted cooling to near-surface temperatures and the attendant pressure release from the cooling, cavity filling occurs. Agates, then, being a product of precipitation of solids from super-critical fluids. The hydrothermal system is then subject to attendant protracted cooling; changing from a hydrothermal system during cycles of eruption followed by cooling periods or periods of erosion.

There are three main types of agates: "volcanic" agate, hydrothermal vein agates, and sedimentary agates.

Hydrothermal vein agates are found in breccia cracks and fissures of the host rocks, usually in volcanics. They form from a mineralized hydrothermal solution and are often related with ore deposits. Comb-like, crustiform, and zonal or growth lines in quartz crystals indicate a direct crystallization from hydrothermal fluids under epithermal conditions. Other hydrothermal minerals are present, such as carbonates, sulfates, and fluorite.

Sedimentary-type agates form in relatively shallow sediments, such as lagoons or a regressive marine shelf environment in saline waters, at surface conditions with temperatures below 40 °C. Silica fills in cavities, vugs, and other empty pore spaces in sedimentary rocks such as limestones, siltstones, argillaceous carbonates, and claystones. Keokuk geodes, for example, found in the Mississippian Warsaw Formation, formed from silicious solutions into syngenetic, diagenetically calcareous concretions. The infilling of the silicious solutions into the carbonate concretions consisted of chalcedony replacing the concretions; the silicification proceeding inward to spherulitic chalcedony and euhedral quartz (Knauth, 1973). Voids are also formed from dissolution of former nodules of soluble material, such as anhydrite, gypsum, or decay of organic material, that dissolves leaving open void spaces. These regressive marine limestones are then exposed to air, creating weathering minerals such as pyrite. The limestone is later covered by air-fall ash, providing the silica source, and thus creating the agate, commonly known as Fortification agate, Dryhead agate, and Fairburn agate. Silicification of concretions and frequent pseudomorphism occurs during sedimentation or early diagenesis of the sediments. Carbonates, sulfates, clays, sulfides, oxides, hydroxides, quartz crystals, celestite, calcite, and barite are often associated minerals. A cauliflower-like surface, lack of horizontal banding, and the presence of quartzine are indicators of sedimentary agates. Quartzine is also a possible indicator of sulfate-rich or evaporitic conditions. Dryhead agate, Fairburn agate, Teepee Canyon agate, Nebraska Blue agate, Mexican Crazy Lace agate, silicified dinosaur bones, and silicified Florida coral are examples of sedimentary agate. The source of silica is thought to be biogenic, coming from diatoms, sponge spicules, and radiolarians, but could be air-fall volcanic ash, as their minerology suggests.

Volcanic agate deposits occur in large igneous provinces in intraplate locations, where the agate deposits are usually associated with continental flood basalt provinces. They are chiefly found in tholeiitic, calc-alkaline igneous provinces, in basalts and andesites (Fig. 3.10). Occasionally they are found in trachytes and latites.

Agates are the products of intense hydrothermal metamorphism, alteration and leaching of tuffs and ashes, at the surface and/or at near-surface depths, involving alkaline waters at temperatures of 20–250 °C. The fluids involve near-surface late phase hydrothermal fluids, and mixed hydrothermal, vadose, and surface meteoric waters. Lebedev (1967), described a silica metacolloid deposit in the andesitic–dacitic volcanics of the Pauzhetka Springs, southern Kamchatka Peninsula, Russia. This is only known instance, in the world, of a silica gel deposit that is a precursor to agate formation. First some background data.

Fig. 3.10 Basalt flow with curved columnar structure, Aldeyjarfoss, Iceland

The Springs discharge from faulted Middle-Late Pleistocene agglomeratic tuffs that are a part of daciandesite and basaltic-andesite lava volcanic sequence, next to the Koshelev and Kambalnyi stratovolcanoes. The sequence is interstratified with dacitic, andesitic, and rhyolitic tuff (pumice) deposits. At the well site, in the borehole, the lithology is described as dacitic-andesitic tuffs. Temperatures at the surface are 90–100 °C. The temperature increases to 170–180 °C at a depth of 800 m. The silica content of the water is 0.2 g/L and the pH is 8.6. Three discrete zones are distinguished where, through alteration and leaching processes, silica is released and precipitated out at the surface, forming a silica gel.

At the surface, at depths of ~5 m, is the kaolinization zone, where the rocks are altered to clays, especially kaolinite and montmorillonite.

At depths of ~30 to ~250 m is the zeolitized zone where andesitic-dacitic tuff glasses and plagioclase have been altered to the calcic zeolite laumontite, with silica released. Zeolitization of the dacite tuffs involved much hydration in the formation of zeolites and the release of silica, enriching the waters in silica. The silica released here is then the silica that forms the agate and thundereggs at the surface or near surface. The calcic zeolites formed are analcime, laumontite, heulandite, wairakite, stellarite, and, rarely, mordenite, and thompsonite and indicate the presence of released silica. Orthoclase as adularia is present as crystals in leached cavities, replacing plagioclase. The dark minerals, chiefly hornblende and augite, show little or no alteration, are slightly chloritized, and are occasionally pyrite replaced as pseudomorphs. Hematite alters to magnetite; ilmenite alters to anatase, brookite, and sphene. Chlorite and celadonite are widespread in this zone, where the celadonite is chiefly pseudomorphic after the tuffaceous dark minerals. It forms in capillary veinlets and as radial aggregates in small cavities.

At ~250 to ~800 m below the surface is the propylitized zone, where alteration and leaching are the most intense: principally, the glasses and dark minerals are replaced by chlorite; the plagioclase has been replaced by calcite and anhydrite; hematite has been replaced by magnetite and pyrite; with silica deposited in the form of chalcedony as vein and cavity filling hydrothermal chalcedony, where it is found in the upper parts of the propylitized zone in the cavities and veins and is associated with quartz, calcite, anhydrite, chlorite, and pyrite. It is less common in the zeolite zone and is found at the surface in botryoidal or collomorphic form as agate.

The silica gel found at the surface is amorphous, isotropic, has globules and chain aggregates, membranous tubes, and has opaline layers. Colloidal or metacolloidal, botryoidal (reniform aggregate), and stalactitic (flow-deposited) textures are found, indicating mineral formation from a gel.

Opal is present at the surface as sinter and is widespread. Quartz is present as quartz crystals of several generations of growth and is prevalent in the upper propylitized zone.

Weathering and alteration of surficial rhyolitic rocks, such as ignimbrites, volcanic ash, tuffs, and volcaniclastic rocks (Fig. 3.11), promotes the formation of agatized thundereggs that are formed in situ within the cavities of a rhyolitic, welded ash-flow tuff. The altered ash or weakly indurated tuff in the rhyolite flow is exposed to late state hydric metasomatic waters or to silica-rich meteoric waters; quite possibly, to an interaction of both water types. The heated hypogene (ascending) and/or meteoric waters release silica from unstable volcanic glass. However, structural features of the thunderegg such as horizontal banding, faulting, and sediment influx and deposition suggest a large involvement of meteoric waters. Chihuahuan geodes and Oregon thundereggs frequently have silt deposited in rock cavities with attendant sedimentary structures, such as graded-bedding, cross-bedding, and slump structures. This seems to indicate periodic circulation of vadose groundwater. The sediments found within the thunderegg nodules or geodes seem to indicate that their formation is a surface phenomenon.

According to oxygen isotope and fluid inclusion studies of volcanic agates, the temperature of agate formation is widespread, both at high temperatures (>200 °C) and at low temperatures (20–100 °C) in a sedimentary and diagenetic environment. High temperatures would indicate mobilization and accumulation of silica during a late phase of, or soon after, volcanic activity. Participation of residual magmatic fluids and the heating of meteoric fluids would be necessary for the alteration of the rocks, silica release, and secondary mineral formation. Fluid inclusion studies indicate a mixing of late magmatic waters with smaller amounts of meteoric water. $\delta^{18}O$ values confirm fluid mixing. The heated waters alter the volcanic rock and release silica. The presence of paragenetic minerals such as carbonates, clays, zeolites, and Fe compounds also indicates that volcanic agate is associated with late or post-volcanic alteration processes within the host rock and underlying rocks. The silica

Fig. 3.11 Weathering of volcanic ash, West Texas

source coming from the alteration of calcic plagioclase feldspars, glass, olivine, and clinopyroxene, among others.

In many agate localities, however, the mineralogy, textures, and the broad areal extent of agates, with the absence of hydrothermal alteration indicators, indicate low temperatures of agate formation, with the alteration of volcanic glass involving low-temperature mobilization, transport, and re-deposition of silica as H_4SiO_4 in the cavities of the host rock. The alteration process of the volcanic ash or weakly indurated tuff involves the formation of phyllosilicate minerals, principally, illite, kaolinite, montmorillonite, smectite, and bentonite through either hydric metaso-matism (the argillic alteration of volcanic glass by low-temperature hydration) or a low-temperature chemical reaction with meteoric waters in epigenetic conditions (Holzhey, 2001) or a combination of both processes.

Another source of silica is derived from the alteration of feldspars in arkosic aren-ites, feldspathic sandstones, and volcaniclastic sediments, the latter also including glassy rocks. For example, the origin of the silica for Brazilian agates suggests that the melting of sandstone xenoliths in flood basalts provided the source of silica (Streider & Heeman, 2006). Another example is the formation of Australian opals from the intense leaching of volcaniclastic sediments with the argillic alteration of orthoclase feldspar to kaolinite and attendant leaching of silica: $KAlSi_3O_8 + H_2O \rightarrow K^+ + H_4Al_2SiO_9 + SiO_2$.

The basaltic-andesitic host rock, itself, can be a source of silica. With lava contact with surface waters or heated meteoric waters, the basalt-andesite is altered from calcic plagioclase to chalcedony and clays (illite, montmorillonite, or kaolinite); the pyroxenes to hematite/goethite/limonite, and chlorite; lime sediments/limestone, if present, to calcite. In addition, alteration of the host rock by hydrothermal waters can be one of the sources of the alumina-silicate zeolites.

Agates are indicators of unconformities through the erosion, weathering, and alteration of the host and source rocks. The iron oxides found indicate pre- or syngenetic formation, and occur at the interface of host rock, in the agate, and accumulate in bands in the agate. The environment consists of arid, alkaline playa lakes as indicated by the evaporitic minerals found associated with the agate. Sandstones and conglomerates indicate braided stream deposits, alluvial fan complexes, flood-plain deposits, and channel deposits that are the sedimentary deposits that indicate water reworking of volcaniclastic sediments, volcanic ash, and ash-flow tuffs that overlie the host rock.

3.20   Diagenesis of Silica

Silica is transported and accumulated as "mobility by metastability" (Ostwald, 1896) with the diffusion and accumulation of H_4SiO_4 molecules in pore solutions. The metastable forms of silica are formed with a sluggish ripening process converting amorphous silica to opal to chalcedony to crystalline quartz.

The inorganic precipitation of silica in natural environments cannot be understood completely in terms of known experimental relationships. In addition, the effects of the time factor involving thousands and millions of years are unknown. There are no present-day comparisons. The geological problem is similar to the problem of chert formation and dolomitization genesis. Silica goes into a protracted heating and cooling history, with cycles of eruption, erosion, and weathering, with silica precipitating from alkaline fluids. Silica, then, is formed in solution as monomeric silicic acid H_4SiO_4. Above a pH of 9 and with high temperatures (100–140 mg/L@20 °C and 300–380 mg/L@90 °C), the extensive ionization of silicic acid increases enormously, with the attendant increased solubility of all forms of silica.

Later low-grade hydrothermal activity, with low temperatures ranging from 50 to 200 °C, with circulation of meteoric waters at shallow depths, with or without a magmatic heating component, and cycles of eruption followed by erosion, heating, and cooling of the silica gels, starts the process of crystallization and banding of the agates. Air bubbles created by boiling are evident in some agates. Replacement textures such as lattice blades and pseudo-acicular and micro-textures such as feathery, flamboyant, mosaic, and jig-saw-puzzle are good indicators of boiling conditions in an epithermal environment.

Amygdaloidal agate seems to form from a diffusion process via the intergranular pore space of the host rock initiated by the pressure reduction caused by the erosional removal of overburden. Due to the large size (1–100 nm) and low diffusion rates of colloidal silica, the main transport method is via diffusion of monomeric silicic acid, H_4SiO_4, to the cavities or vesicles of the host rock. Through a condensation process, the silicic acid flocculates, and transitions from a silicious sol, to a gel, and ending up as an amorphous silica precipitate. Many colloidal textures, such as reniform

Fig. 3.12 Botryoidal and membraneous tube texture in agate

aggregate, membraneous tube, stalactitical, botryoidal (Fig. 3.12), and other plastic deformation types, corroborate the colloidal nature of agates (Lebedev, 1967).

Precipitation of silica occurs through either evaporation, introduction of electrolytes (weathered or altered minerals, such as Al and Fe hydroxides from clays), and supersaturation of the water. Precipitation can also occur through the reduction of pH, of alkaline, silica-rich waters passing through carbonate deposits, causing CO_3 to dissolve and silica to precipitate.

In volcanic silica deposits, agates seem to form at higher temperatures at the opal-CT/C stage, without the earlier opal-A dissolution phase present in the formation of chert. The formation of chert has the initial opaline gel converting to crystalline quartz (see section on chert).

The multiple stage diagenesis of silica has been recognized as a series of complex dissolution–precipitation events. The diagenetic sequence is characterized by increasing crystallinity, crystal size, and structural order. The diagenesis of silica proceeds as silica in solution as H_4SiO_4, to a sol, then to a gel, then to an amorphous precipitate, to Opal-A, to Opal CT, to chalcedony, granular microcrystalline quartz, and finally to mega quartz. When the silica content of the precipitate is low, the sequence is short-circuited and mega quartz or quartz crystals form.

Moxon (2002) and Rios et al. (2023) found a decreasing amount of silanol water with increasing geological age. The total water content varied in the agates between 0.23 and 1.78 wt% and the average amount of structural silanol water was found to be 76% of the total water content. They concluded that silanol groups were released during the recrystallization of moganite to chalcedony, and therefore an estimation of the age of agates based on the amount of structural water should be possible.

As the total water content decreases as microstructures become more ordered and compacted, mineral densities increase as they become more crystalline (Lee, 2007). With the increase in age and diagenesis, the resulting improved crystallinity XRD patterns show sharpening and narrowing of reflections with increasing age of agates. Further, if one plots agate crystallite size versus log age of the host rock, there is a general increase in crystallite size with increasing age of the host rock (Moxon, 2002). This correlation between agate crystallite size and age of the host rock suggests that the formation of wall-banded agate is penecontemporaneous with the formation of the host rock (Moxon, 2002).

3.21 Agate Banding

It was originally thought that the banding in agates was a rhythmic influx of silica to the voids and cavities of the host rock. Liesegang (1915), after experimenting with silver nitrate in a potassium dichromate gel in a silica gel, that formed concentric rings similar to agate banding, observed that the formation of rhythmic color bands in agate was due to a diffusion process of metal ions in silica gels. Banding or collomorphic texture is common to most colloidal or metacolloidal deposits.

Two types of banding are evident in agates. The first type is the wall-lining agate or the so-called "fortification" agate. The bands conform to the outline of the host rock cavity. The other type is the horizontally layered agate or the "Uruguay" or "water-line" agate; the layering caused by the gravitational sedimentation of sols with flocculated SiO_2 particles. Recent SEM work (Götze, 2009) discovered that the white bands are in actuality plate-like structures that are able to totally reflect incident light, while the non-white (gray-blue) areas allow some white light to be transmitted. What this means concerning layering in agates is unclear.

Banding seems to be is due to the variations of the silica phases and their impurities. The most common impurity is red hematite (Fig. 3.13). Crystallization forms layers in the agate according to the silica and impurity content of the precipitate. The precipitate crystallization proceeds from a high silica content, twisted, spherulitic chalcedony layer with a high impurity content to a lower silica content phase and layer of macrocrystalline quartz, producing a crystallization front or band that becomes more and more morphologically stable with a characteristic drop in silica content, impurities, and structural defects (Wang & Merino, 1990). With the silica precipitate negatively charged and the impurities positively charged, the impurities precipitate or migrate into bands (Zarins, 1977). The compaction of the outer bands near the outer surface of the nodule and the diapiric deformation of the bands are flow structures that show the agate was still in a gelatinous state after the banding formed. Plumes, moss, and sagenitic structures are the beginning stages of agate banding and, if given enough time and heating, produce bands. Banding that is in large number and is finely defined, as in the Lake Superior agate, seems to indicate agates of great age.

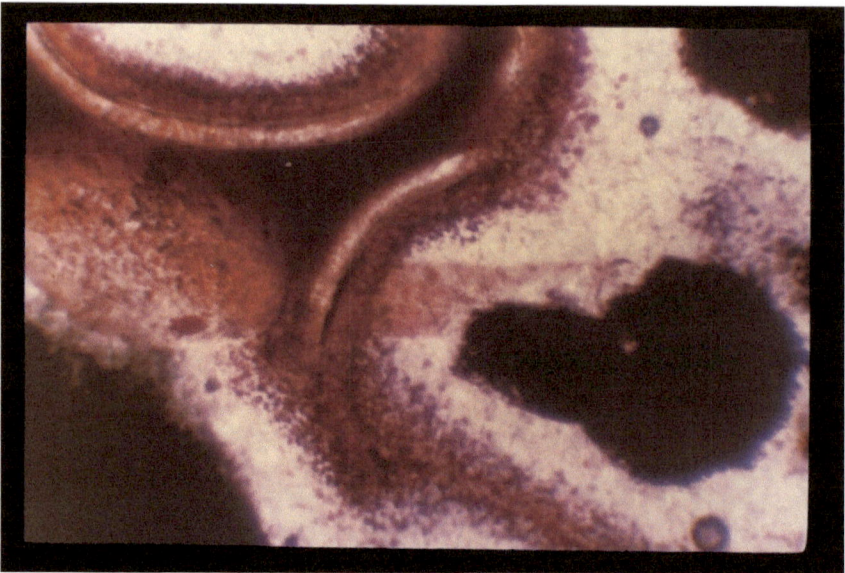

Fig. 3.13 Globules of hematite precipitating into bands in red carnelian; ×23, plane light

3.22 BZ Reaction

Minerals derived from weathering or alteration minerals, if present, react through the BZ reaction process, with the silica gel, producing plumes and bands. The BZ reaction is a chemical non-equilibrium reaction that produces non-linear chemical oscillation patterns and is an example of non-equilibrium thermodynamics; it is an incomplete or partial chemical reaction that is never completed (Zhabotinsky, 227). The Belousov–Zhabotinsky reaction or BZ reaction takes place with the weathered/altered minerals (hematite, limonite, goethite, pyrolusite, psilomelane, and chlorite) within the silica gel, producing the agate bands, plumes, and moss dendrites. The manganese minerals are very soluble and are flushed out of the agate, leaving a white residue. The alkaline, carbonate, and evaporitic mineral suite, consisting of calcite, aragonite, magnesite, and siderite, have the same electrical charge as the silica gel, and do not react with the gel, forming crystals, crystal molds, and pseudomorphs within and outside of the agate. The phyllosilicate clays, illite, chlorite, and celadonite generally form on the skin of the agate and are the initial first layer of the agate. Visit this site for a visual presentation of the reaction: YouTube:

- Reaction—Variety 1
- Reaction—Variety 2
- Reaction—Variety 3
- Reaction—Variety 4.

3.23 Agate Coloring

Agate color is due to variations in crystal size, microstructure, porosity, and water content. Agate with no colored banding and no inclusions or plumes indicates a lack of hydrothermal, weathering/alteration, and evaporitic minerals within the lavas, with the resultant lack of chemical impurities in the agate. White, gray, opaque, milky white, and blue-gray colors, i.e., the colorless bands, are the most dominant colors found in agate, and indicate closely packed chalcedony fibers, with small, water-filled cavities. The blue color is produced by the scattering of light, termed Rayleigh scattering.

Other colors are due to the impurities in the precipitate. Black, red, orange, and brown colors are indicative of the presence of iron minerals and are the most frequent colors found. A lavender color is caused by iron inclusions in blue chalcedony; a green color, by chlorite. Iris agates are a rainbow effect produced by transparent thin slices of finely banded agate, consisting of narrowly spaced layers of chalcedony, each with a different index of refraction.

3.24 Conduit Canals

Ductile deformation zones of agate bands are thought to be either escape or infiltration tubes for silicic fluids (Fig. 3.14). Noggerath (1850) postulated that the canals are entrance canals for the admission of percolating fluids. Bauer (1904) thought that the canals were tubes of entry or escape due to rhythmic deposition of silica laden, hot springs, or geysers. However, not all agates have canals.

Shaub (1955) demonstrated that the feature is an escape tube by showing that more than one such feature could be found in an agate nodule that they had no preferred orientation, and that some did not extend all the way to the surface of the nodule.

The compaction of the outer bands near the outer surface of the nodule and the diapiric deformation of the bands seem to be flow structures that show the agate was still in a gelatinous or plastic state after the banding formed, with most of the flow tubes formed by the deformation of the non-solidified silica in a plastic state. As pressure increases during the formation process, parts of the banding are squeezed outward. As crystallization from the silica gel continues, the agate, having a lower specific gravity, occupies a larger volume; hence, material is squeezed from inside the agate nodule toward the outside, the bands forming an inverted V shape (Pabian & Zarins, 1991).

Fig. 3.14 Conduit canal in agate

3.25   Paragenetic Minerals and Other Mineral Associations

Paragenetic minerals are found in volcanic agates as mineral inclusions in quartz, intergrowths with quartz, or pseudomorphs (Fig. 3.15). Chemical elements maybe present in different chemical compounds in the same agate and indicate highly variable physio-chemical conditions, especially redox conditions. The mineral associations present in volcanic agate indicate formation under low-temperature hydrothermal conditions. Calcite, aragonite, hematite, limonite, goethite, pyrolusite, psilomelane, and, occasionally, magnesite are the corresponding paragenetic minerals found in volcanic agates.

Secondary zeolites, along with calcite and the clay minerals, celadonite, smectite, illite, and chlorite, formed and deposited in vesicles and joints from the alteration of volcanic glasses and host rock pyroxenes, and plagioclase, in stages, from late phase hydrothermal magmatic fluids that involved meteoric waters in the later stages (Ottens, 2019). They are found especially associated with copper deposits, formed and deposited at very low temperatures, at approximately <60 °C, as is the case with the Keweenaw copper deposits. The more common zeolites deposited include natrolite, thompsonite, mordenite, powellite, pavellite, heulandite, and apophyllite.

Paragenetic calcite has a ubiquitous occurrence in agates in the forms of intergrowths, crystals, and pseudomorphs. Several carbonate generations can be present. Oxygen isotope studies of calcite indicate the mixing of at least two different fluids with temperatures ranging from 20 to 230 °C.

Later host rock weathering and alteration in an oxidizing environment, also, produces hematite, limonite, goethite, psilomelane, pyrolusite, and chlorite. The minerals occur at the interface of host rock in vesicles, joints, and fractures, and

Fig. 3.15 Chalcedony pseudomorph after aragonite

accumulate as crystals and bands in the agate and are the product of weathering of the host rock as opposed to the hydrothermal precipitation of these minerals. In the more basic host rocks, celadonite forms between the agate and host rock. The zeolite, clinoptilolite, and clay minerals, especially bentonite, form from the alteration of tuffs and ashes.

In addition to the paragenetic mineral suite and the weathering and alteration mineral suite, an evaporitic mineral suite is also found, indicating a sabkha-like, arid, alkaline environment with playa lakes, that formed on the lava surfaces. Aragonite (orthorhombic, pure $CaCO_3$) precipitates or evaporates out of warm waters at temperatures of 80–100 °F. Calcite (rhombohedral, $CaCO_3$ with some magnesium) forms an isomorphous series with magnesite ($MgCO_3$) and ferroan dolomite, at a high pH, will rapidly precipitate out along with magnesite, $MgCO_3$, and siderite, $FeCO_3$. They are found frequently as chalcedony pseudomorphs after the evaporitic mineral, indicating that colloidal silica formed from the weathering of silica-rich sources such as volcanic ash and tuffs in the arid, alkaline environment (Fig. 3.11). The minerals in this suite are calcite (also as vadose carbonate nodules), aragonite, dolomite, siderite, gypsum, anhydrite, barite, pyrolusite, and magnesite.

Minerals found associated or adsorbed with agates are presented in Table 3.1.

Table 3.1 Minerals associated with agates

Chemical class	Mineral
Elements	Copper
Oxides/ hydroxides	Quartz opal hematite goethite limonite pyrolusite psilomelane ilmenite magnetite
Carbonates	Calcite aragonite dolomite siderite magnesite
Sulfates	Barite celestite anhydrite gypsum
Sulfides	Pyrite
Silicates	Adularia albite clay minerals (illite montmorillonite, smectite, kaolinite, celadonite, chlorite, bentonite)
Silicates (Zeolites)	Laumontite clinoptilolite natrolite thompsonite apophyllite, mordenite powellite heulandite

3.26 Agate Formation Time Frame

The time frame for formation of agate is thus initially in hundreds or thousands of years for the formation of the colloid and the initial BZ reactions; millions of years for subsequent crystallization and diagenesis of the agate. Stratigraphic and age-dates of various agate provinces suggest that the agatization process is on the scale of 1–2 million years. Thus, the mineralization and the diagenetic patterns of chalcedony and quartz within the agate reflect the erosional, weathering, heating, and cooling history of the surrounding volcanic deposits.

3.27 Thermodynamic Laws for Agate Formation

Some hypothetical viewpoints speculated as "thermodynamic laws" of agate forma-tion: (1) the first law is the Ostwald ripening and formation of larger crystals on smaller crystals in a sol or agate with smaller colloidal particles accumulating around larger nuclei; (2) the second law is the BZ chemical reaction in agates; and (3) the third law is specific heat needed for the formation of agate during diagenesis and crystallization.

Chapter 4
Chert

There are two schools of thought on formation of chert. The first theory believes that chert formed from biogenic sources in oceans, namely, from silica secreting organisms such as foraminifera, radiolarians, and diatoms. They supposedly use all the soluble silica supplied by rivers and streams that empty into the oceans. It is argued that these organisms only use about 10% of the soluble silica. Thus, the formula: biogenic opal → opal-CT → chert (granular microcrystalline quartz). The second theory states that chert formed from silicates of clays and zeolites that had interacted previously with clays and bicarbonates. Thus, the formula in inorganic precipitation: silica in solution → amorphous silica → opal-CT → chalcedony/ granular microcrystalline quartz → megaquartz. However, due to under-saturation in present-day oceans, silica cannot precipitate inorganically. Chert occurs in limestone as nodules, sheets, lenses, and beds. Black chert is flint.

The Interlocking anhedral grains of α quartz are 2–30 μ in size, have undulatory extinction, and diffuse grain boundaries (Fig. 4.1). Chert consists of 0.27% H_2O with 1% OH water that is bonded to silicon. The quartz grains are built up on piles of α-quartz ∥ (perpendicular) to the (001) axis. The faces of the quartz grains are covered with Si–OH groups and water. Water is an indication of alternating plates of right- and left-handed quartz. Twinning occurs according to Brazil's law. Crystals are formed from epitaxial replacement of consolidated silica gel—a subsidiary cryptocrystalline silica consisting of disordered growth of cristobalite and tridymite. The occurrence of opal and the high-water content between opal and microcrystalline quartz suggest that chert is a diagenetic precipitate derived from the dissolution of amorphous silica.

A. Zarins, *The Geology of Agate Deposits*,
SpringerBriefs in Earth System Sciences, https://doi.org/10.1007/978-3-031-67929-2_4

Fig. 4.1 Thin section of chert, "salt and pepper" texture; polarized light

4.1 Diagenesis of Chert

At low temperatures, Silica-A or Opal-A is precipitated out, inorganically or organically, from natural aqueous solutions. Biogenic Opal-A is formed from the dissolution of organisms that are diatoms, radiolarians, and siliceous sponges (Fig. 4.2). The high surface areas of the organisms give rise to high solubilities and rates of solution that lead to supersaturation and the precipitation of siliceous oozes.

Chert has carbonate fossils, relict bedding planes, dolomite rhombs, cavity filling, veins, chalcedony, and alpha quartz, rarely opal and opal-CT. Nodular cherts are secondary, replacing diagenetic limestone, but is penecontemporaneous with carbonate sedimentation; mostly from biogenic sources. Novaculites such as the Caballos Novaculite of Arkansas and Texas formed from biogenic sources.

Volcanic cherts, on the other hand, are non-fossiliferous and are formed from hydrothermal waters in areas of volcanic activity such as island arcs, vents, rift valleys, geothermal areas, continental hot spots, transform faults, and spreading lineaments. Their fabrics or textures are of seven types and include the equigranular microcrystalline texture (the same as in sedimentary cherts) and fibrous textures, such as spherulitic chalcedony, zebraic chalcedony, and micro-flamboyant quartz (Hesse, 1989). But here we are entering the realm of agates with their spherulitic growth textures, etc. Volcanic cherts contain altered ash and glass material, indicating the silica source. Stained red by iron oxides and referred to as jasper and stained blue-green-black by ferrous (reducing) iron, they constitute the Banded Iron Formations

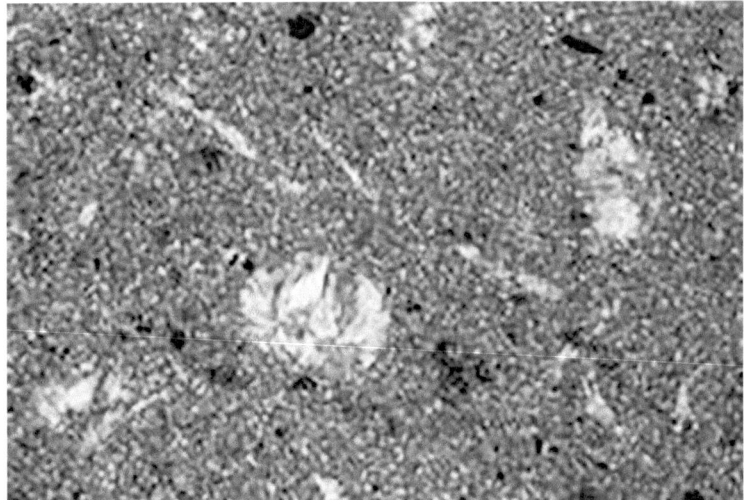

Fig. 4.2 Thin section of chert with radiolarian and sponge spicule fossils, plane light

(BIFs) that are alternating bands of chert, jasper, and iron ore. BIFs produce 90% of the world's supply of iron.

A variety of chert that forms by decalcification is tripolitic chert. It forms by the removal of groundwater, of carbonate grains, that are mostly bioclastic, included within the chert. It is not a common variety, as in most cherts, the carbonate content is very low. Lower Mississippian Upper Boone formation, Mississippi Lime, and the Arkansas Novaculite cherts are examples of a chert that does contain enough fossil carbonate grains to be leached.

Another variety of cherts are the inorganic cherts formed in alkaline lakes. They are initially deposited as sodium silicates. Percolating and circulating ground water removes the sodium and the mass is then converted to chert and chalcedony.

4.2 Chert Paleotemperatures

The $^{18}O/^{16}O$ ratio of marine cherts of central and western US was calculated by Knauth (1973). The temperature of chert formation is 20°–34 °C for Paleozoic cherts and 35–40 °C for Triassic cherts. The temperatures decreased to a present-day value of 13°–15 °C. Precambrian temperatures are over 50 °C, indicating much warmer ocean temperatures. Also, these temperatures reflect differing amounts of meteoric waters and the presence of meteoric waters. Further, the $^{18}O/^{16}O$ ratio changes across diagenetic boundaries in chert, with abrupt decreases in ^{18}O across opal-A/opal-CT/ quartz boundaries, implying higher temperatures are the catalyst that causes the higher forms of crystallization.

4.3   Some Interesting Thoughts About Chert

The chert iron formations, BIFs, are almost exclusively pre-Cambrian and are >2 billion years in age. The carbonate chert deposits formed from 2 billion years ago to 60 million years ago, with a maximum formation in the late Paleozoic. The mid-continent Mississippian carbonates of the United States are especially prolific in chert. From the end of the Mesozoic to the present, chert is formed exclusively from diatoms, utilizing all dissolved silica in the oceans.

Chapter 5
Thundereggs

Thundereggs are small spheroidal chalcedony bodies about 3 inches in diameter with a cauliflower-like surface crossed by low ridges (Fig. 5.1). Most thundereggs consist of an outer shell of pale-brown aphanitic rock that is composed of volcanic glass shards, fine ash, and collapsed pumice lapilli; all of which are altered to radially oriented sheaves of fibrous cristobalite and alkali feldspar. The core consists of white-to-bluish-gray chalcedony that contains concentric and planar bands and dendritic mineral growths. The cores range in shape from round and highly irregular forms to geometrically regular pyritohedrons and cubes; each face of which is an inward-pointing pyramid that appears as squares and stars in section.

5.1 John Day Formation, Oregon

Thundereggs are found in the base of the late Oligocene to Early Miocene John Day Formation, south of Pony Butte, Oregon, in a rhyolitic ash flow, termed member F (Fig. 5.2). The basal portion of the Member F is a poorly indurated tuff and lapilli tuff with an underlying weakly welded ash flow. The tuff and lapilli tuff are colored in shades of green, yellow, red, and gray; they are mostly massive ash-fall deposits, but include some bedded water-laid tuff. The basal ash flow is 5–25 ft thick and is fairly well indurated, and is light gray speckled with abundant chips and blocks of black glass as much as 4 inches in diameter. In the basal 4–6 inches of the ash flow, fragments are progressively flattened downward and the rock is moderately to strongly welded.

Ross (1941) explained the origin of the thundereggs found there. The spherulites were formed during cooling of the rhyolitic ash flow. The spherulites were disrupted by the pressure of volatiles exsolved from the ash. The resultant cavities were later filled by silica gel during alteration of the enclosing ash. Cavities may or may not have been filled completely at once or may have been filled rhythmically or periodically

A. Zarins, *The Geology of Agate Deposits*,
SpringerBriefs in Earth System Sciences, https://doi.org/10.1007/978-3-031-67929-2_5

Fig. 5.1 Thunderegg, John Day Formation, Oregon

as evidenced by planar layering of chalcedony. Thunderegg formation is considered an in situ process.

However, it has been determined by Holzhey (2001) and others (American Minerologist, 2013) that the lithophysae were formed not by exsolved gases, but by water under pressure in the pores of the perthite and cristobalite spherulites, turning to steam, and thus creating the star-like cavities. Further, the lithophysae were formed by potassic, sodic, and hydric metasomatism. Hydrogen metasomatism is the last stage of metasomatism according to Holzhey that involves silica mobilization from the silicified spherulites and form the chalcedony, agate, and macro-quartz within the cavity. According to Holzhey, there is interaction of primary and meteoric waters at this stage, with scalenohedral calcite considered to be co-genetic with the deposition of silica. The chalcedony isotope compositions of agates and calcite indicate low temperatures of formation (ca. 70–130 °C). Secondary mineralization consists of the formation of alteration minerals of illite, chlorite, calcite, and quartz.

Like other thunderegg deposits, thundereggs are found exclusively in rhyolites and ignimbrites in the glassy component of the ignimbrite or rhyolite flow, in the spherulites or lithophysae. The lithophysae or spherulites develop from the cooling of a highly siliceous melt that has a high volatile gas content and has a similar geochemistry to the rhyolite/ignimbrite rocks and lithophysae. The rhyolite and spherulites are usually altered from sanidine, orthoclase, and quartz to kaolinite, illite, and smectite clays. The mineralogy of the rhyolites and spherulites indicates a high temperature of formation accompanied by rapid cooling. Götze (2023) fluid inclusion studies of the Saxony thunderegg indicate temperatures of formation from 134 to 186 °C with the mobilization and accumulation of silica into the spherulites or lithophysae during a

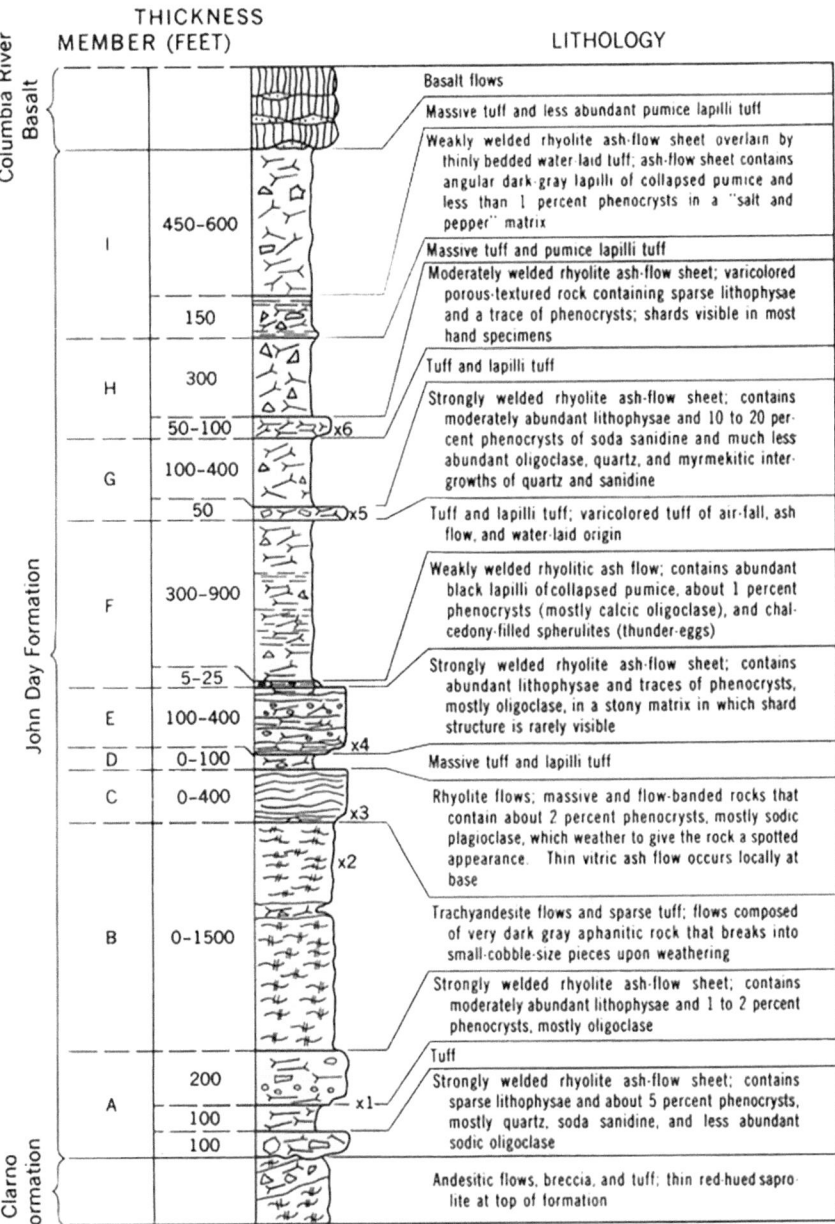

THICKNESS
MEMBER (FEET)   LITHOLOGY

Columbia River Basalt

Basalt flows

Massive tuff and less abundant pumice lapilli tuff

I   450-600   Weakly welded rhyolite ash-flow sheet overlain by thinly bedded water-laid tuff; ash-flow sheet contains angular dark-gray lapilli of collapsed pumice and less than 1 percent phenocrysts in a "salt and pepper" matrix

Massive tuff and pumice lapilli tuff

150   Moderately welded rhyolite ash-flow sheet; varicolored porous-textured rock containing sparse lithophysae and a trace of phenocrysts; shards visible in most hand specimens

Tuff and lapilli tuff

H   300

50-100   x6   Strongly welded rhyolite ash-flow sheet; contains moderately abundant lithophysae and 10 to 20 percent phenocrysts of soda sanidine and much less abundant oligoclase, quartz, and myrmekitic intergrowths of quartz and sanidine

G   100-400

50   x5   Tuff and lapilli tuff; varicolored tuff of air-fall, ash flow, and water-laid origin

F   300-900   Weakly welded rhyolitic ash flow; contains abundant black lapilli of collapsed pumice, about 1 percent phenocrysts (mostly calcic oligoclase), and chalcedony-filled spherulites (thunder-eggs)

5-25   Strongly welded rhyolite ash-flow sheet; contains abundant lithophysae and traces of phenocrysts, mostly oligoclase, in a stony matrix in which shard structure is rarely visible

E   100-400

D   0-100   x4

Massive tuff and lapilli tuff

C   0-400   x3   Rhyolite flows; massive and flow-banded rocks that contain about 2 percent phenocrysts, mostly sodic plagioclase, which weather to give the rock a spotted appearance. Thin vitric ash flow occurs locally at base

x2

B   0-1500   Trachyandesite flows and sparse tuff; flows composed of very dark gray aphanitic rock that breaks into small-cobble-size pieces upon weathering

Strongly welded rhyolite ash-flow sheet; contains moderately abundant lithophysae and 1 to 2 percent phenocrysts, mostly oligoclase

Tuff

A   200

100   x1   Strongly welded rhyolite ash-flow sheet; contains sparse lithophysae and about 5 percent phenocrysts, mostly quartz, soda sanidine, and less abundant sodic oligoclase

100

Clarno Formation

Andesitic flows, breccia, and tuff; thin red-hued saprolite at top of formation

John Day Formation

Fig. 5.2 Stratigraphic column of the Priday thunderegg beds, Oregon (Peck, 1964)

late volcanic or hydrothermal activity phase or soon after. Further, rare earth element (REE) and heavy rare earth element (HREE) indicate interactions of the host rocks (rhyolites and ignimbrites) with SiO_2, magmatic volatiles (F, Cl, and CO_2), and meteoric and surface waters. Cathodoluminescence (CL) shows a characteristic yellow CL. Internal textures and a high defect density of macro- and micro-quartz detected by electron paramagnetic resonance (EPR) spectroscopy indicated crystallization from an amorphous silica precursor under non-equilibrium conditions.

Further, Götze's paper explains that pseudomorphs after chalcedony of chiefly carbonates and sulfates indicate secondary alteration and replacement by silica, the carbonate and sulfate minerals being products of primary hydrothermal fluids. The spherulitic growth occurs from an amorphous silica predecessor that seeps into the lithophysae from small fissures and fractures in the host rock. A discontinuous supply of silica is indicated by the crystalline layers or "fronts." Lesser concentrations of silica develop macro-quartz and any impurities in the gel accumulate in the center of the agate or thunderegg.

With the interaction of heated meteoric waters and hydrothermal fluids that dissolved the overlying and surrounding volcanic ash and tuffs to clays and silica-rich solutions, chalcedony and agate infillings were solidified and hardened in the lithophysae cavities, forming the thundereggs.

The 10–20' thick basal member of member F of the John Day formation reflects this alteration process, as the basal member is altered from a black perlitic, angular lapilli of collapsed pumice in a matrix of shards and ash, to clay and opal, locally. Most of the thundereggs are found in this layer. The horizontal layering of the chalcedony found in many of the thundereggs reflects fluctuating meteoric groundwater movement and elevations. In addition, many of the thundereggs also show mini-faults zones and sediment influx or deposition.

5.2 Deming, New Mexico

Another well-known thunderegg locality is at the Rockhound State Park, southeast of Deming, New Mexico, in the Little Florida Mountains, where extensive andesitic to rhyolitic volcanism from middle Eocene to early Miocene times formed the Little Florida Mountains (Fig. 5.3). This volcanism consisted of ash-flow tuffs, air-fall tuffs, flow-banded rhyolite, basaltic andesite, and dacite. Thick rhyolitic fanglomerates in the Little Florida Mountains and alluvial conglomerates formed an apron around the mountains and were deposited as the mountain block was up lifted approximately 7,000 ft since early Miocene times (Clemons, 1998).

600 ft above the east parking lot, above a foot path, is an outcrop of perlitic rhyolite that contains stretched lithophysae (Fig. 5.4). The rhyolite flow is altered and interbedded with rhyolitic tuffs. The tuffs and the rhyolite flow are age-dated to the Oligocene at 24.4 My and the rhyolite flow. A silicified fanglomerate overlies the rhyolite flow and is Miocene in age.

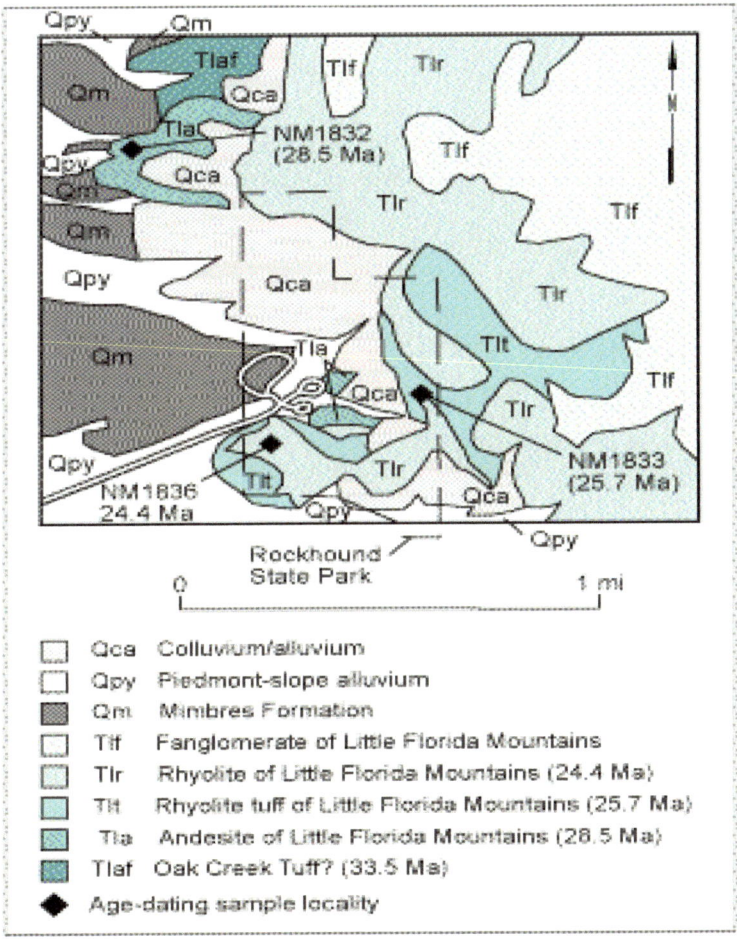

Fig. 5.3 Geologic map of Rockhound State Park, Little Florida Mountains, New Mexico (McLemore & Dunbar, 2000)

The thundereggs are found in the altered rhyolite flow. The rhyolite ranges from grayish pink and pale red to dark grayish red, is hypo-crystalline to microcrystalline, and contains less than 1% phenocrysts. Flow-banding is prominent locally near the margins of the intrusions, but generally the rock is massive; an autobreccia texture predominates. The breccia in the southeast end of the range contains prolific cavities, apparently caused by alteration and removal of angular autoliths; some cavities up to 30 cm across. Some cavities contain spherulites that formed after removal of the clasts, indicating potassic metasomatism. Spherulites up to several inches in diameter are common near contacts at the southern end of the Little Florida Mountains. The spherulites formed from the nucleation, coalescence, and expansion of vapor bubbles and consist of spherulitic intergrowths of quartz, alkali feldspar, plagioclase, and

Fig. 5.4 Thunderegg locality, Rockhound State Park, Little Florida Mountains, New Mexico; thundereggs are in the rhyolite flow in center of picture

magnetite (McLemore & Dunbar, 2000). Some of the spherulites have white to bluish-gray chalcedony centers. The chalcedony is derived from late-stage siliceous-rich, inter-mixed magmatic fluids and groundwater that deposited the silica within the cavities. A feathery texture is common in the thundereggs and is an indication of boiling fluids. In some of the thundereggs, an angular unconformity is present between the banded agate and the horizontally layered agate indicating landslides or tilting of local fault blocks. Some of the spherulites ruptured or broke apart. The spherulite is then later infilled with chalcedony veins; an indication of a separate later phase fluid, separate and apart from the spherulite growth phase. In addition, calcite, that is also found in some of the thundereggs, indicates precipitation from another separate fluid with different physical and chemical conditions, involving the loss of CO_2 due to pressure loss and a rapid increase in temperature. It is estimated that the thundereggs formed from temperatures <200–340 °C.

Geodes, partly filled with chalcedony and occasional vugs containing clear quartz crystals, are common in a zone extending from Rockhound State Park southeastward across the range. Moderately reddish-brown jasper fills fractures in the rhyolite and tuff near the contacts throughout this same zone.

The western and northern parts of the Little Florida Mountains are composed mostly of the flow-banded rhyolite, with associated obsidian, rhyolitic tuff, and tuffaceous beds. An ash-flow tuff, dated 32–37 Mya, showed hydrothermal alteration of plagioclase to clay, indicating the release of silica. At least half, and maybe most, of the rhyolite exposed along the western escarpment and crest of the Little Florida Mountains as elongate, irregular, domal to dike-like intrusions. Their contacts are

obscure, but appear to range from vertical to steep, east or north-east-dipping atti-
tudes. Many of the intrusions between Rock Hound State Park and the southeast
end of the range have black, perlitic obsidian border zones. These rocks intruded
penecontemporaneous, greenish- to orangish-gray lithic tuffs and volcaniclastic beds
deposited around the vents.

Vesicular to amygdaloidal basaltic-andesite flows overlie the rhyolite, altered tuff,
and obsidian northeast of Mamie Canyon near the south end of the Little Florida
Mountains. The flows are overlain by massive fanglomerate breccias. There is a
small outcrop of the basaltic andesite near a manganese prospect on the west side of
Mamie Canyon. The dark grayish-red to reddish-brown basaltic andesite is composed
of altered plagioclase laths in an interstitial matrix of iron-oxide grains and brownish
cryptocrystalline to holohyaline material. A chemical analysis indicated probable
potassium metasomatism of this rock. The interior vesicles are filled mostly with
calcite and some chalcedony. The exact age of the basaltic andesite is tentatively
considered to be early Miocene (post-23.6 My).

Chapter 6
Petrified Wood

Petrified wood is the replacement of wood tissues by silica by a sequential silica diagenesis process: colloidal silica → opal-A → opal-CT → chalcedony → macrocrystalline quartz (Mustoe & Dillhoff, 2022). It is technically a quartz pseudomorph after wood. The chalcedony phase includes agatized petrified wood. Multiple silica phases in a single specimen indicate either multiple episodes of mineralization or the process of silica diagenesis. Interestingly, the diagenetic process seems to be related to the age of the petrified wood, as most Paleozoic and Mesozoic petrified wood is primarily cryptocrystalline or microcrystalline quartz, whereas Cenozoic petrified wood is mainly opal (Mustoe, 2015). Initially, wood is enclosed in a volcanic ash/breccia that was deposited in a fluviatile sand or sandstone or a mud or shale environment. The water is saturated with silica, dissolved from volcanic ash, and is in a high pH or alkaline state. The waters are generally meteoric involving surface water, connate water, and groundwater. The decaying wood caused the pH to drop and silica to precipitate out.

Silica molecules, derived from silica sources such as volcanic ash, penetrate the vascular tissue in plants. The silica molecules form hydrogen bonds with polysaccharides and other carbon-based molecules present in plant cellulose and lignin. As original carbon tissue degrades, additional hydroxyl molecules become available for bonding with polymerized silica (Dayvault, 2005). Then, the silica diagenetic polymorphs are formed, e.g., Opal-A to Opal-CT to chalcedony.

Petrified wood is formed by cell replacement of the cell walls and open spaces of the wood by amorphous silica gels and maybe analogous to geode formation, where multiple episodes of hydrothermal precipitation and layering occur. Gradual mineralization, or permineralization, occurs initially in cell walls, in tissues that are also experiencing microbial degradation (Mustoe, 2015). Permineralization is the process where a large amount of the original woody material is retained as silica (or calcite, siderite, apatite, pyrite) fills pore spaces and cavities of the wood. A large amount of the original material is retained. Replacement is the other process that rarely forms alone but occurs with permineralization. The dissolved silica in

A. Zarins, *The Geology of Agate Deposits*,
SpringerBriefs in Earth System Sciences, https://doi.org/10.1007/978-3-031-67929-2_6

water dissolves the original material and is replaced by silica, replicating the original structure of the wood. Cell walls are progressively dissolved. The cellulose and lignin are degraded and replaced by silica. The slower the rate of replacement, the more detailed and defined are the original tissues of the wood. The aqueous solutions have a reduced content of the amount of oxygen and thus reduces the deterioration of organic material by fungi and molds. Wood that is buried in an anaerobic environment, such as mud, is protected from microbial degradation and is preserved in an unmineralized carbonized state.

6.1   Petrified Forest, Arizona

The geologic setting—The Pangea supercontinent is beginning to break up at the beginning of the Triassic. The petrified forest, itself, is located in east central Arizona. The petrified forest is found in the Late Triassic Chinle Formation (Fig. 6.1). The Chinle Formation consists of continental deposits of major river systems that are lowland terrestrial deposits of river channels, floodplains, swamps, and small lakes in an alkaline environment.

The rocks suggest a strongly seasonal subtropical climate with an up-section transition to an increasingly arid climate. Iron, manganese oxides, and carbon in the water provide the colors of the petrified, agatized wood to yellow, red, black, blue, brown, white, and pink (Fig. 6.2).

The major occurrence of the petrified wood is in the Sonsela member, and is locally abundant in the Petrified Forest member and Blue Mesa members. The rocks are principally, sandstone, claystones, and mudstones with pedogenic carbonate nodules of paleosols common. The Sonsela sandstone has altered volcanic pebbles, a varicolored bentonitic claystone. The sediments are volcanic in origin that relict volcanic textures and altered tuff grains and are highly montmorillonitic, the montmorillonite formed by alteration of tuffs. Most volcanic debris—especially volcanic glass—is highly unstable under near-surface conditions. Its chemical composition is such that it readily—almost spontaneously alters to montmorillonite and silica (Schultz, 1963). The tuffs are altered after deposition by water. The rounded particles suggest transport by water in an alkaline fluvial or lake environment. The type of clays indicates the parent rock type: mixed illite/montmorillonite clays are altered from potassium-rich rhyolite tuffs, and montmorillonite clay (especially in the Petrified Forest member) is altered from a potassium-poor latitic tuff. The Sonsela formation is very montmorillonitic.

The petrified wood is from at least nine species of trees, all of which are now extinct. Araucarioxylar arizonicum, Woodworthia arizonica, and Schilderia adamanica are the most common species found. Also, lycopods, ferns, cycads, conifers, and ginkgoes are found. Fossil vertebrates found include the phytosaur reptiles; the earliest combination dinosaur and amphibian.

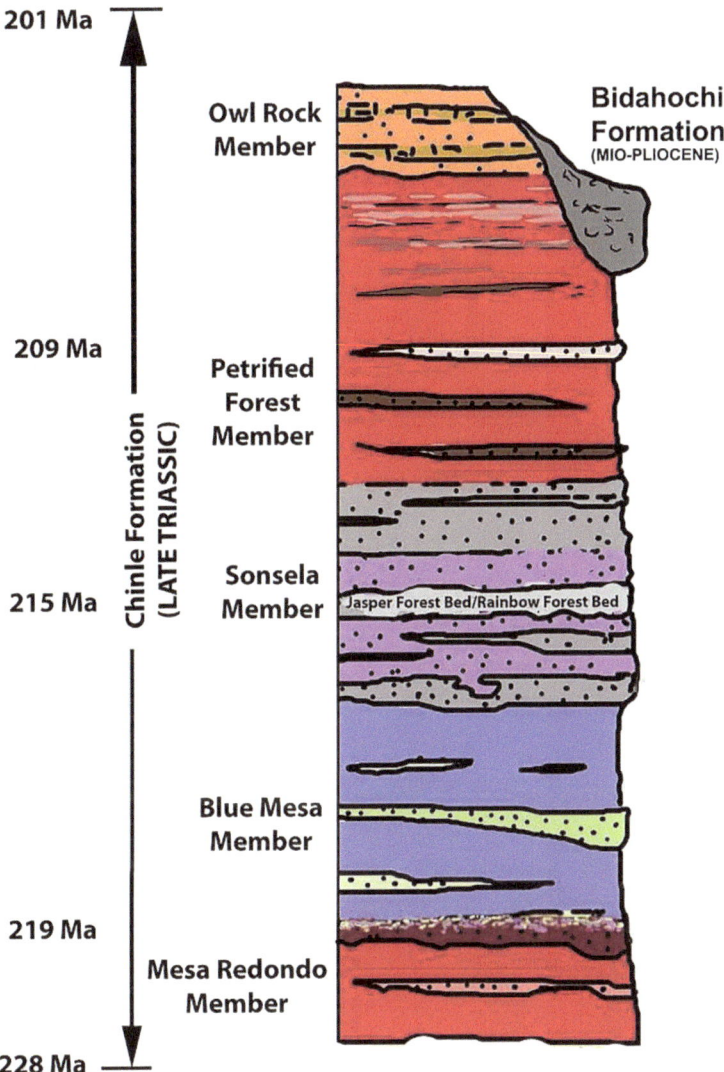

Fig. 6.1 Stratigraphic column for the Petrified Forest (Martz et al., 2012 and National Park Service/Public Domain)

Rate of petrification is unknown. For your consideration, radio isotopic dates place the late Triassic Chinle Formation between 225.2 and 209.9 Mya. The Sonsela formation was deposited in 10 My.

Fig. 6.2   Petrified logs in the Petrified Forest, Arizona

Chapter 7
Volcanic Agate Provinces

7.1 Luna County, New Mexico

Agate occurs in Luna County, New Mexico at several localities around the Deming area (Griswold, 1961). The best occurrence and most well-known deposits are in the Burdick Hills (Fig. 7.1). The deposit lies in secs. 10, 11, 14, and Hermanas Road, State Highway 331, and thence 5 miles due east on a dirt road.

The rock outcrops at the agate deposits are all of volcanic origin. The volcanics are divisible into two units: an upper or "caprock" unit of Tertiary gray latite, and a lower unit of light-gray, altered tuff. The agate is restricted to the tuff beds. The latite "caprock" is barren, forming a layer ranging from 0 to 50 ft thick. The structure of the deposit suggests a dome, in which case the curved outcrop of latite south of the main outcrop may represent a second layer of latite that originally overlay the deposit but was removed by erosion. The Gila (?) conglomerate is exposed west of the agate deposit. The agate occurs both as small vein infillings and as small pods in the tuff. The veinlets are randomly oriented, generally of very short length, and extremely variable in width. The better agate is well banded and possesses any combination of the following colors: colorless, white, gray, blue-gray, brown, and reddish brown. However, the most prevalent appearance is clear or colorless. Moss agate is also present. Occasionally thin, flat-lying lenses of agate are found near the surface. These lenses may represent either an original agate zone that formed at the top of the tuff beds immediately below the latite "caprock," or they may have been formed from vein agate that accumulated residually at the surface during erosion. Most probably, free silica was formed from the weathering and alteration of the tuff beds. The silica then accumulated as vein infilling and as pods in the cavities of the underlying or unaltered tuffs.

Interestingly, a thunderegg deposit (the Hermanas deposit) is found nearby about 1 mile northwest of Hermanas. The thunderegg agate occurs in altered latite flows and tuffs. The thundereggs are almost round nodules ranging from 1 inch to almost 1 foot in diameter. The "eggs" generally have an outer zone of mottled-brown spherulites

SpringerBriefs in Earth System Sciences, https://doi.org/10.1007/978-3-031-67929-2_7

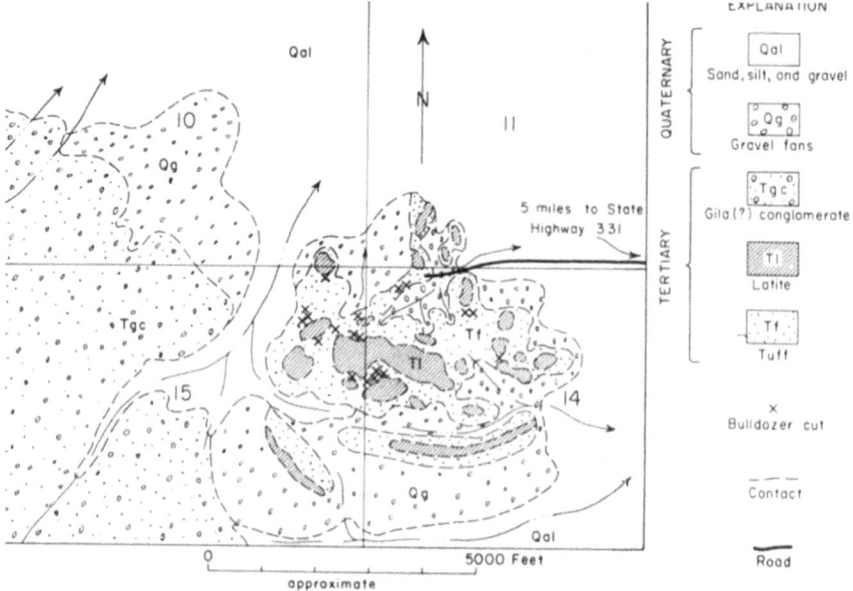

Fig. 7.1 Geologic map of the Burdick Hills agate deposit, Deming, New Mexico (Griswold, 1961)

and an inner zone of blue-gray chalcedony or agate with an occasional center of quartz crystals.

7.2   Cathedral Mountain West Texas

West Texas plume agate is found adjacent to Cathedral Mountain, south of Alpine, Brewster County, Texas on the Woodward Ranch. The field is located on the southern edge of the Davis Mountains (Figs. 7.2 and 7.3). The Mexican highlands, the Basin and Range physiographic provinces, and the Appalachian and Laramide orogenic belts all intersect here. This is where the collision and rifting of the African and South American plates took place; the plates subducting under the North American plate in late Paleozoic to Early Mesozoic times. It is part of the failed Rio Grande rift zone and could be termed an aulacogen.

7.2.1   Host Rock Stratigraphy and Lithology

The site is located in a calc-alkaline volcanic province. The volcanics are part of the Tertiary Buck Hill Volcanic Series, with a thickness that approaches 4,600 ft. The volcanic pile erupted on the Cretaceous Boquillas limestone. The formations are

Fig. 7.2 Regional map of the Trans-Pecos Texas Volcanic Field; Cathedral Mountain Quadrangle is southeast of Paisano Pass Caldera (Davidson, 2014)

from oldest to youngest: Pruett Tuff, Crossen Trachyte, Sheep Canyon Basalt, Potato Hill Andesite, Cottonwood Spring Basalt, Duff Formation, Mitchell Mesa Rhyolite, Tascotal Tuff, and Rawls Basalt (Fig. 7.4). An intrusive alkalic microsyenite forms sills, dikes, laccoliths, and a trap-door dome in the area.

Age determinations and geochronology of the Trans Pecos Texas volcanic field were determined by Goldich (1949), McAnulty (1955), Ross and Dietrich (1970), Wilson (1980), Duex et al. (1981), and more recently by Davidson (2014). The relevant age determinations are as follows. Vertebrate fossils indicate a late Eocene (Duchesne) age for the Lower Pruett Tuff. The Crossen Trachyte has a K–Ar date of 37.8 Ma and is considered Early Oligocene. The overlying lavas and tuffs are Oligocene and younger (?) in age. The Paisano Volcano, north and north-west of Cathedral Mountain, and source of the Duff Formation volcanics, has K–Ar dates of 35 Ma—Middle Oligocene, and indicates that it was an active volcano for 1–2 Ma. The intrusives are younger than the lavas. It has been observed that the Crossen Trachyte gains appreciable quartz northward; the Potato Hill Andesite changes from highly porphyritic rock to non-porphyritic rock; and that the Duff Formation changes

Fig. 7.3 Satellite photo of the Cathedral Mountain Quadrangle, south of Alpine, Texas; Calamity Creek fault in center of photo (Google Earth, 2022)

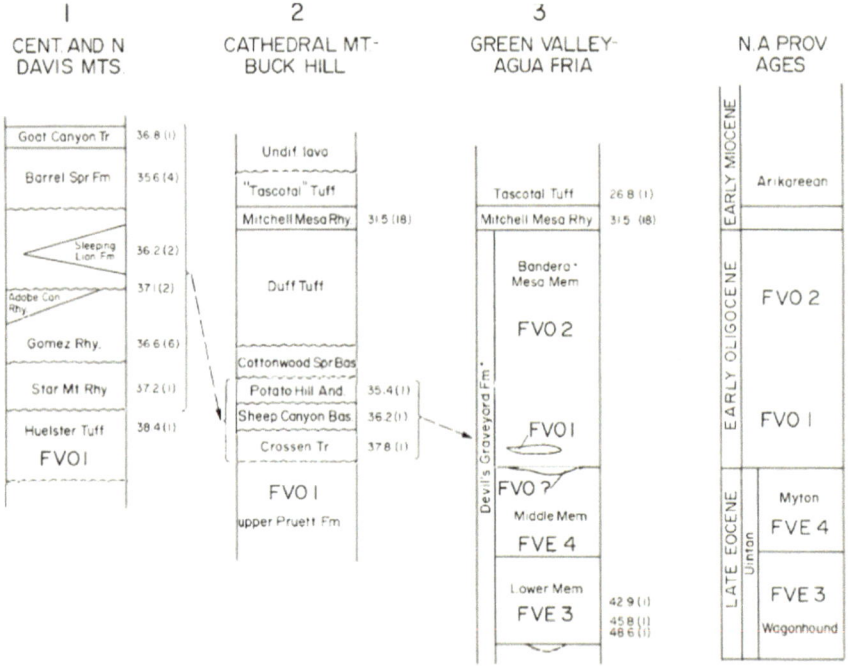

Fig. 7.4 Stratigraphic column and regional correlation of the Buck Hill Volcanic series (Wilson, 1980)

from a dominantly rhyolitic tuff rock to a lava, tuff, and conglomerate (Decie member) rock (McAnulty, 1955).

7.2.2   Crossen Trachyte

The Crossen Trachyte contains numerous vesicular zones; with elongated vesicles trending northwesterly to southeasterly. There is no Crossen trachyte present between the Pruett tuff and Sheep Canyon Basalt in the southeastern part of the quadrangle, indicating pre-Sheep Canyon erosion, creating a disconformity. The trachyte consists of alkali feldspar with phenocrysts of anorthoclase and sanidine, with some corrosion and alteration to carbonate, hematite, and chlorite. There are sporadic phenocrysts of altered aegirine-augite, with some accessory magnetite and apatite. Primary quartz can be as high as 15%. In which case, the rock becomes a rhyolite. The high silica content of 71% is explained by the presence of introduced secondary chalcedony and quartz.

7.2.3   Sheep Canyon Basalt

The Sheep Canyon Basalt is a series of basaltic lavas, intercalated by thin limestone and tuff beds. The number of lava flows varies from section to section; anywhere from 3 to 8 flows are identifiable. Vesicular tops and bottoms are common. The vesicular tops are green-colored, chloritized zones formed during late stage, deuteric, (low temperature) alteration. Fragments of unaltered basalt are common and are incorporated into the chloritic groundmass. Hematite and magnetite are abundant.

The deuteric or diagenetic alteration of the basalt to hydrous minerals, such as chlorite in this case, indicates that the basalts have undergone extensive isotopic exchange with low temperature meteoric waters, by using δD and $\delta^{18}O$ isotope analysis (Livnat, 1976). Fine-grained to massive-bedded siliceous limestones and tuffaceous limestones with siliceous fossil fragments and algal structures occur at an erosional disconformity between the Crossen Trachyte (where up to 150 ft of Crossen Trachyte was removed locally); and within the Sheep Canyon Basalt flows indicating that basalt flows invaded developing lacustrine sedimentary sequences. The basalt flows flowed into water bodies, such as lakes, and were subsequently covered by the lacustrine deposits. Numerous thin tuff beds, also between the basalt flows, have variable compositions: some strongly silicified and indurated, others baked and compact, while others are friable, white-gray alteration product. Analysis of the tuffs indicated the presence of montmorillonite, alpha quartz, and clinoptilolite, indicating a rock rhyolitic in composition. The basalt, itself, is a dark green to black, fine to medium grained, basalt with phenocrysts of labradorite as the primary mineral. Aegirine-augite and titano-augite are the main pyroxene minerals. Olivine crystals show some alteration to antigorite and chlorite. The accessory minerals are magnetite and apatite.

Alkali feldspars occur interstitially in small quantities, sometimes as high as 10%, in which case it is classified as a trachybasalt.

7.2.4   Potato Hill Andesite

The Potato Hill andesite is a distinctive aa flow with an upper flow-breccia member and a lower massive, very vesicular flow member. A thin indurated tuff generally separates the units. Irregular, vesicular fragments are in the flow breccia. The andesite weathers in thin platy sheets; in some localities, the flow is weathered to a red residual soil. Weathered surfaces are red-brown with fresh surfaces a steel-gray color. Outcrops are fine-grained with phenocrysts of plagioclase. Phenocrysts disappear to the north, and the andesite becomes aphanitic. The massive member is trachytic to granular, with predominantly andesine plagioclase present as small laths. Some alkali feldspar is present. Accessory minerals include magnetite, hematite, olivine, iddingsite, antigorite, and chlorite.

The brecciated member is similar to the massive member but shows more oxidation. Large volumes of chalcedony and calcite cement the brecciated member. The high silica content accounted for by large amounts of introduced chalcedony and quartz. A large percentage of hematite reflects the extensive and intensive oxidation.

7.2.5   Cottonwood Spring Basalt

The Cottonwood Spring Basalt is a purplish gray basalt with a speckled schistose appearance due to the presence of plagioclase feldspar laths (labradorite), arranged in the direction of the lava flow. Many vesicular and amygdaloidal zones are present in numerous flows. There are at least four flows. Flow breccias are common, with the basalt up to 332 ft thick. The basalt is a Pahoehoe lava that flows from the north and northwest, thickening to the north. It is highly dissected with differential erosion: spheroidal weathering in gullies and low areas and platy splitting on cliffs and steeper areas. It weathers to a reddish-brown; highly weathered condition. It caps many of the small hills where erosion has stripped the Duff Tuff away (Fig. 7.5). Incomplete erosional remnants measure from a few feet to 332 ft. The only complete section of the basalt is at Cienega Mountain, where it is 286 ft thick and difficult to correlate. The mineralogy consists of orthoclase that mantles some of the labradorite. The alkali feldspar is sometimes abundant as the calcic feldspars but generally is in smaller percentages than calcic feldspar. The main mafic mineral is augite. Some olivine, (generally altered to iddingsite, antigorite, and chrysotile), analcime, magnetite, and apatite are the other accessory minerals found. Much of the rock is weathered and hematite obscures much of the mineralogy. Calcite and chalcedony occur as secondary minerals filling vesicles. Minor zeolitization, occurring as stilbite, occurs in one locality at North Branch Walnut Draw.

Fig. 7.5 Cottonwood
Springs Basalt Agate Field,
Woodward Ranch; Eagle
Peak Vent to the left;
Cathedral Mountain in
distance

7.2.6   Duff Tuff

The Duff Tuff overlies the Cottonwood Springs Basalt. It originates from the Paisano
Peak Caldera, north of the agate fields (Fig. 7.2). Two lava flows (the Decie Member
of the Duff Tuff) occur in the middle to upper Duff Tuff; one, a porphyritic trachyte
and the other a rhyolite porphyry. These flows thin to the south around Cathedral
Mountain. The zeolites, analcime, natrolite, and thompsonite, as does calcite, occur
as amygdules in vesicles and in weathered, moist areas of fissures and cracks of the
flows. Secondary alteration of primary minerals in the flows is fairly severe.

Laterally the flows change to a tuffaceous, sandstone-boulder conglomerate
(Fig. 7.6), indicating flood-plain and stream channel deposits, poorly sorted, with
scour and fill features, cross-bedding, and rapid facies changes. Airfall tuffs and ash
flows interfinger with the sandstone-conglomerate member; crudely bedded with
some re-working. Progressive devitrification of the airfall tuffs is indicated with the
formation of cristobalite, alkali feldspar, and biotite (Fig. 7.7). The glass shards are
75–95% iron-stained. The tuff is a silica-rich, rhyolitic tuff.

Fig. 7.6 Duff Tuff
Conglomerate Member

Fig. 7.7 Duff Tuff, airfall
tuff member, with glass
shards and alkali feldspar
phenocrysts in an
iron-stained matrix, ×60,
plane light

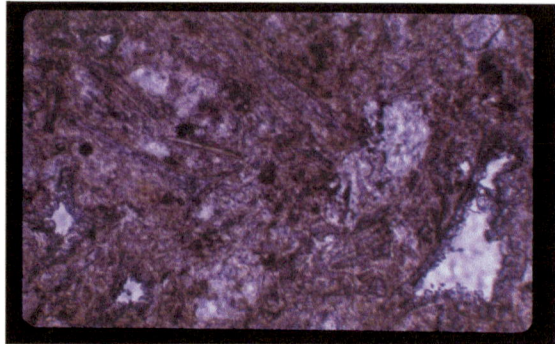

Alteration and re-working of sediments are evident where the tuffs interfinger with the sandstone-conglomerate member. Glass shards are altered to clays and zeolites. Hematite and magnetite microcrystals are present. Basaltic clasts and unaltered feldspar are also present.

The sandstone-conglomerate member consists of 30–40% basaltic clasts, that are rounded to angular, and enclosed in a hematitic groundmass. Importantly, no agate or chalcedony clasts are found within the conglomerate, indicating that agatization is post-Duff tuff. No agate or chalcedony is found in the Decie member lavas, also indicating that agatization is post-Duff tuff.

7.2.7   Mitchell Mesa Rhyolite

The Mitchell Mesa Rhyolite is a pink to gray rhyolitic ash flow tuff that erupted 32 Mya from the Chinati Mountains, 55 miles southwest of Cathedral Mountain (Fig. 7.2), where the best exposure of the rhyolite is found (Duex et al., 1981). Cathedral Mountain is an erosional remnant that is uplifted by an intrusive syenite. The stratigraphy suggests the rhyolite eruption formed the Chinati Mountains Caldera.

7.2.8   Tascotal Tuff

The Tascotal Tuff is a light gray to yellow, ash flow tuff, interbedded with sandstones and conglomerates. It erupted 31.2 Mya. Paleocurrent directions indicate the Chinati Mountains Caldera as the probable source area. There is an alteration of the ash and tuffs to clays, opal, calcite, clinoptilolite, and hematite. Fresh volcanic shards are found in the eolian member of the formation.

7.2.9 "Rawls" (Petan?) Basalt

The Rawls Basalt consists of very fine-grained, porphyritic flows of trachybasalt with flow breccia zones. The basalt is age dated to 26.2–17 Ma, making the basalt Miocene in age. The source area is unknown but is probably the Chinati Mountains.

7.2.10 Intrusive Syenite

The syenite intrudes the Buck Hill Volcanic Series, trending northwest to southeast, as sills, domes, laccoliths, and plugs. It up-lifted Cathedral Mountain. The rock is an alkalic micro-syenite with the dominant mineral an alkali feldspar. Aegirine-augite is the main mafic mineral, with some altered olivine, hematite, sericite, calcite, and present.

7.2.11 Eagle Peak Vent Agglomerate

The agglomerate is associated with the Cathedral Mountain plug (Fig. 7.5). It is a well-cemented, fine-grained volcanic breccia, composed of small to medium lapilli of basalt and andesitic lava and fragments of baked tuff. No silicification is present. The agglomerate could be evidence of the formation of later extrusives in the area.

7.2.12 Structural Geology

A normal fault, the Calamity Creek Fault, runs on the east side of the agatized areas. The faulting is post-Duff tuff, probably caused by the intrusion of the syenites. It is a primary pathway for ascending hydrothermal waters.

7.2.13 Agatization and Origin of Agates in the Cathedral Mountain Quadrangle

Consequent stream development and erosion took place on a broad volcanic field extending to the east and southeast of the Cathedral Mountain quadrangle. The amount of erosion that took place since the emplacement of the Rawls basalt is considerable. Goat Mountain section measurements indicated that the Duff Tuff is 1,399 ft thick, Mitchell Mesa Rhyolite is 90 ft thick, Tascotal Tuff is 190 ft thick, and the Rawls Basalt is 210 ft thick, indicating a total of at least 1,889 ft of formation erosion in the Cathedral Mountain Quadrangle since Post-Cottonwood Springs Basalt (Early Oligocene) times.

Further, considering the erosion of post-Cottonwood Springs basalt and the Duff tuff: the Crossen Trachyte was emplaced 37.8 Mya and the Duff Tuff was emplaced 35 Mya; indicating 2.8 million years of erosion of the Cottonwood Springs Basalt is possible. The Duff Tuff was emplaced 35 Mya and the Mitchell Mesa Rhyolite, further up-section, was emplaced 32 Mya; making for a possible 3 million years of erosion of the Duff Tuff. The Tascotal Tuff was emplaced 31.2 Mya; making for 0.8 million years of erosion of the Mitchell Mesa Rhyolite. The Rawls Basalt was emplaced 26.2 Mya; making for 5 million years of erosion of the Tascotal Tuff. The last flow of the Rawls basalt is 17 Mya; making for 17 million years of erosion to date.

The large volume of agate requires a large volume of source material. The Woodward Ranch Cottonwood Spring basalt plume agate dimensions are approximately 150 ft thick, 1,300 ft wide, 4,700 ft in length; the basalt contains approximately 8% agate. This equals 73,320,000 cubic feet of agate alone.

The agates of the Cathedral Mountain quadrangle were formed from siliceous gels originating from the dissolution of the overlying Duff Tuff volcanic ash or tuffs that formed siliceous lakes or ponds, primarily on top of the Cottonwood Springs Basalt, the Sheep Canyon Basalt, and the Potato Hill Andesite and is a near-surface or surface phenomena. The location of the agatized areas occurs in a northwest–southeast band or swath adjacent to the Calamity Creek Fault, suggesting that late phase hydrothermal fluids migrated from the fault, dissolved the overlying Duff Tuff or ash, and deposited silica directly on the basalts and andesites. The zeolites, analcime, natrolite, and thompsonite, formed from the alteration of subsurface rocks, were deposited and are found in the Decie Member of the overlying Duff Tuff, are evidence of hydrothermal waters passing through the Duff Tuff: silica was precipitated out from late-stage hydrothermal waters, that was probably mixed with meteoric waters, forming colloidal solutions in an alkaline environment, in temperatures at the surface, as low as ambient surface temperatures, but probably more in the range of 50–120 °C, or 90–100 °C. The temperatures in the subsurface, in the zeolite and propylitized zones, were much higher, probably in the 170–180 °C range (Lebedev, 1976).

The likely source of the silica required to form the agates of the area is probably the tuffaceous units that occur on the agatized volcanic rocks. The field evidence supports this hypothesis. The Duff Tuff is silica-rich; tuffaceous sandstone-boulder conglomerates occur between the lava member and the airfall tuff member of the Duff Tuff. The sandstone-conglomerate member shows scour and fill structures, poorly sorted beds, rapid facies changes, and cross-bedding indicative of stream and floodplain deposits. Considerable re-working of sediments and erosion had occurred in a three-million-year period. No agate or chalcedony is found in the Duff tuff lavas, sandstones, or conglomerates; indicating that agatization must be post-Duff Tuff, and hence could not be attributable to any pre-Duff Tuff igneous activity.

Silica-rich solutions infiltrate the underlying rocks, accumulating in vesicles, cracks, fissures, and joints permeating the rock. Permeation of silica is proved by the chemical analysis of the Crossen Trachyte and Potato Hill Andesite, which have an abnormally high percentage of silica for a trachyte and andesite due to introduced chalcedony. Precipitation is accomplished by evaporation, supersaturation, or contact with electrolytes. The electrolyte, Fe_2O_3, that constitutes the plumes and color of the agate, must have been in solution with the silica or was introduced from the vesicle walls of the rock, being a weathering product of the basalts/andesites. In addition, calcite precipitates out at high pH and is frequently found in the center voids of the agate or tops of lenses of agate, but not within the agate or gel. Chalcedony pseudomorphs after aragonite indicate an original high alkaline environment where aragonite formed from the evaporation of alkaline solutions at temperatures of 80–100 °F. Silica gels have considerable electrical charges and attract iron salts. Once in the vesicle, the filling probably expands as silica particles change from elongate to globular, also inhibiting escape from the vesicle. The colloidal textures (reniform aggregate, globular, microcollomorphic, membranous tube, stalactitical, spherulitic, and botryoidal) are further evidence of growth from colloidal solutions.

The stratigraphic evidence indicates that agatization is post-Duff Tuff or Oligocene in age. The Duff Tuff surface eroded and weathered for approximately 3 Ma. Later deposition of the Mitchell Mesa Rhyolite at 32 Ma and the Tascotal Tuff at 31.2 Ma produced a major heating event, and produced similar diagenetic and lithologic characteristics as the Duff Tuff. Another phase of erosion and silica diagenesis took place, lasting 5 Ma, before emplacement of the final volcanic phase and heating event of the Rawls Basalt, ending 17.6 Mya in Miocene times.

7.2.14 Agate Stratigraphy

Various varieties of agate in the Cathedral Mountain Quadrangle area and surrounding areas seem to indicate that the agates could be reliable indicators of erosional unconformities. In addition, the type or variety of agate indicates a particular stratigraphic interval in which they are found.

No evidence of age differential has been detected. It is suggested, however, that the rhythmic banded blue-gray agate of the Sheep Canyon Basalt, reflects at least three major heating events and two major periods of weathering and erosion. The Potato Hill agate has no banding and little plume formation, but, rather, a diffuse coloring. In terms of diagenesis of agate, the Potato Hill agate is a very young or immature agate. The Cottonwood Springs Basalt plume agate, with the lack of rhythmic banding, suggests a younger agate, reflecting one major period of erosion and weathering. The agate types may also reflect a change in the rate of weathering and erosion and/ or a change in the chemistry of the silica-bearing solutions.

7.2.15   Crossen Trachyte Agate

The Crossen Trachyte agate is not a major agate type found within the area. The chalcedony is pink, with or without bands. The color is due to hematite; where it appears as globules within spherulitic chalcedony. The globules are characteristic of growths from a colloidal solution. Hematite globules have a mean size of 20.36 microns and constitute the red bands of the agates. There is an apparent sequence of chalcedony and quartz crystals, with an inner massive ball of hematite common; the quartz growing in radial fashion around a central nucleus of hematite-enriched chalcedony spherules. Reniform aggregate structures are found sparingly. Fracture lines (desiccation lines?) crisscross some agates.

7.2.16   Sheep Canyon Basalt Agate

The Sheep Canyon Basalt vesicles are filled with intergrown chalcedony and calcite. Pseudomorphs of chalcedony after aragonite in intergrowths of calcite and chalcedony in a spherulitic crystal growth pattern are common. A blue-gray agate occurs in the vesicular tops of flows, having a characteristic lack of hematitic banding and plumes. They occasionally contain magnesite spots and have white and blue bands. The centers are occasionally filled with euhedral quartz crystals or mixtures of quartz and calcite crystals; up to 7 cm in length, and 3 cm in width. Microplume or "flower garden" agate is also found locally within the Sheep Canyon Basalt where it occurs in layers or lenses (Fig. 7.8) and is an alteration product formed from the contact of the lava flow with the alkaline lakes. The basalt lava flows were altered to chalcedony from the plagioclase, hematite, limonite, and goethite, from the ferro-magnesium minerals, and calcite from the lime lake sediments. The plumes consist of black, brown, gold, and red hematite, typically growing vertically up from the bottom of the lens or layer, indicating a vertical flow component of the groundwater table. The Agate Hill location has green chloritic plumes. Many of these agates have reniform and membranous tube structures. A red carnelian agate with abundant secondary calcite of the basal pinacoid variety, and spherulitic aragonite, commonly replaced by agate, commonly occurs as layers and lenses in a chloritic matrix, appearing red laced. In thin section, globules of red precipitated hematite are forming layers.

7.2.17   Potato Hill Andesite Agate

The Potato Hill andesite agate occurs in large vesicles of the andesite and are "potato"-like in appearance. They have broad, diffuse, red ("blood" stone), black, clear, and lemon-yellow bands. Some have inner cavities of drusy quartz. A red plume is found in some varieties. This agate has no banding and little plume formation, but, rather, a

Fig. 7.8 Flower garden
agate lens with calcite in
Sheep Canyon Basalt;
Meriwether Ranch

diffuse coloring; probably a sub-variety of the plume agate, or, in terms of diagenesis of agate, a very young or immature agate. On Neville Haynes property, the potato-like nodules of agate were completely eroded out of the flow and lay on the surface. The agate seems to have formed in partially formed pipe amygdules, which are elongated vesicles, formed by gasses that streamed upward from the bottom of the flow.

7.2.18   Cottonwood Spring Basalt Agate

The Cottonwood Springs Basalt agate occurs entirely in vesicles and is similar to the microplume agate of the Sheep Canyon Basalt (Fig. 7.9). The plumes grow horizontally along the length of the agate, indicating a horizontal flow component of groundwater. The plumes are colored black, brown, gold, red and consist of hematite derived from the weathering of the basalt. Some plumes are so thick as to dominate the agate; others surround the outer layer of the nodule only, suggesting plumes grew from the walls of the vesicles. Many agates consist entirely of euhedral quartz crystals; others are hollow cavities filled with drusy quartz or stalactitic growths of chalcedony; others contain pods of the country rock. Many colloidal textures occur within the agate: reniform aggregate, membranous tube (stalactitic), botryoidal, and other plastic deformation types. Frequent chalcedony pseudomorphs after radially spired aragonite are found.

7.2.19   Rawls Basalt Agate

The upper lava flow has flow breccia zones with botryoidal black chalcedony, occurring in cavities and coatings in veins and stringers. Drusy quartz is frequently present. A black chalcedony found in a basalt (probably the Rawls Basalt) is found at Singleton Ranch, approximately 30 miles west of Cathedral Mountain (Fig. 7.10). This is a

Fig. 7.9 Plume agate from
Cottonwood Springs Basalt

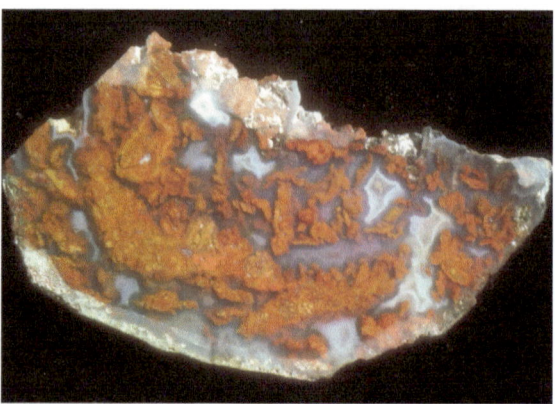

Fig. 7.10 Singleton Ranch
chalcedony field; Rawls
Basalt (?) Over altered ash
unit

black to clear to gray, botryoidal chalcedony, with some red, pink pastel shades within
the chalcedony.

7.3   Lake Superior Agates

With the breakup of Laurentia about 1.1 billion years ago, Early Proterozoic
(Precambrian). Midcontinent rifting began, with two major rifts or arms forming;
one to Kansas and the other to southeast lower Michigan (Fig. 7.11). Extrusion
of lavas began into the rift valleys in the Middle Proterozoic, forming the North
Shore Volcanic Group lavas, chiefly tholeiitic basalts, in Minnesota and the Portage
Lake volcanics, also tholeiitic basalts, with some minor sedimentary rocks, in the
Keweenaw Peninsula, Wisconsin, and Michigan.

The Portage Lake Volcanics consist of about 200 subaerial basalt flows intercalated with clastic sedimentary rocks; a few cm thick to 40 m thick. There are a few
rhyolitic flows and intrusives at the base of the section. The basalts have amygdaloidal

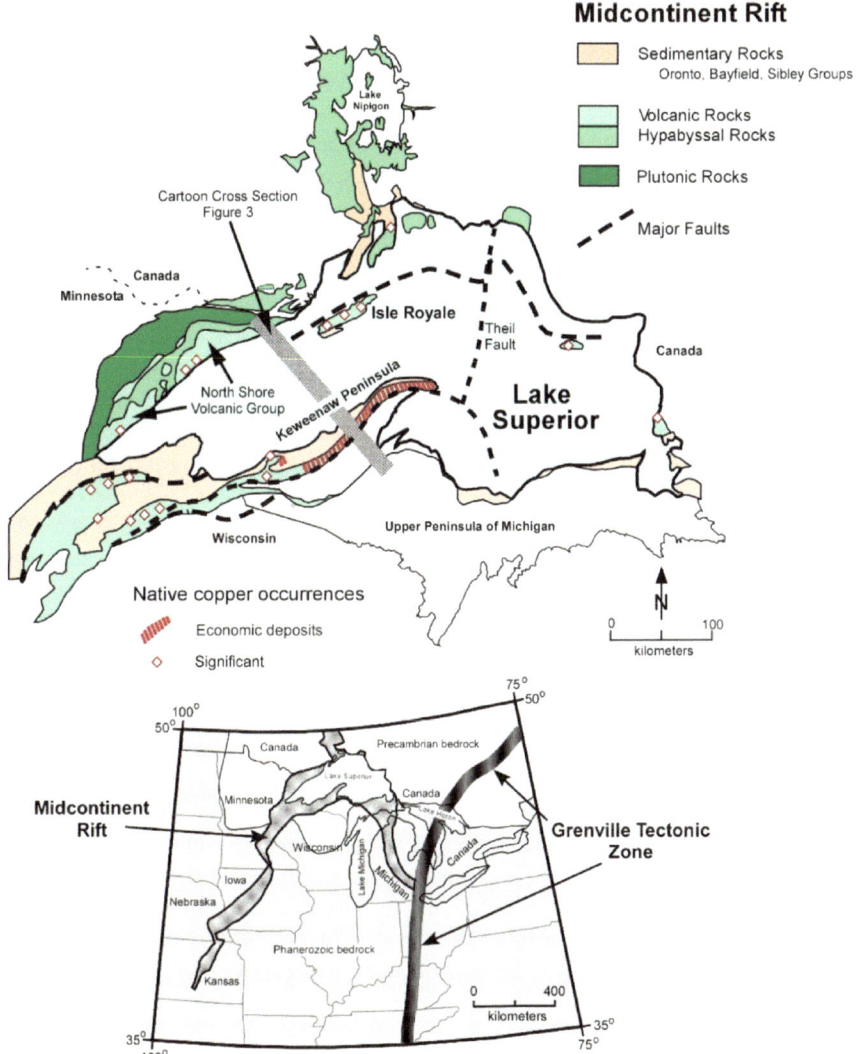

Fig. 7.11 Keweenaw midcontinent rift (Bodden et al., 2022)

or brecciated flow tops and the flows are 10–20 m thick. The brecciated zones are very porous and better host rock for native copper deposits. The sedimentary rocks consist chiefly of a red conglomerate with lesser amounts of sandstone, siltstone, and shale. The conglomerate is pebble to boulder-sized, sub-rounded to angular and consists of predominantly felsic clasts. The Portage Lake volcanics erupted in a time span of 2–3 million years, with the Greenstone flow, in the middle of the volcanic succession, dated to 1.094 ± 1.5 Ga.

The North Shore Volcanic Group in Minnesota occurred about the same time as the Portage Lake Volcanics but is stratigraphically beneath the Portage Lake volcanics and also consists mainly of basalts. Intrusive activity, during this time, formed the Duluth Intrusive Complex.

With the waning of thermal activity at approximately 1.080 Bya, sediments began to fill the rift basins, forming the Copper Harbor Conglomerate. The middle portion of the Copper Harbor Conglomerate includes a succession of lava flows known as the Lake Shore traps (Lane, 1911) that outcrop from Great Sand Bay to Agate Harbor in Eastern Keweenawan Peninsula, the southeastern side of Isle Royale in northern Lake Superior, and Michipicoten Island in eastern Lake Superior. The lowermost mafic flows of the Lake Shore traps were deposited as ponded sheets, while upper andesite flows may have formed a low, topographic, positive feature, such as a shield volcano. The flows have segregation cylinders and numerous sandstone dikes and also have numerous mineralized slickenside surfaces that could reflect the subsidence of the rift sequence. With pahoehoe flow tops, abundant agate, along with lower temperature or metamorphic grade zeolites, are found here. The maximum thickness is 600 m near the tip of the Keweenaw Peninsula where it has 31 flows and one interflow conglomerate. The composition of these lavas ranges from Fe-rich olivine tholeiites at the base to Fe-rich olivine bearing tholeiitic basaltic andesites to tholeiitic andesites at the top of the succession. The lavas date to 1.087 Ga $\pm$ 1.6 Ma from U–Pb age dating of zircon and are the youngest igneous rocks of the Keweenawan Volcanic Series, younger than the Portage Lake Volcanics by 7 million years.

With regional compression, reverse faulting and additional sedimentation occurred, along with native copper mineralization. The copper mineralization or hydrothermal event occurred 1.070–1.040 Bya, with hydrothermal fluids moving along faults and fractures (Bodden et al., 2022). The Keweenaw fault, the Isle Royale fault, and other faults are thought to be conduits for the low temperature hydrothermal fluids that flowed into the high permeability and porosity conglomerates. With deuteric or diagenetic alteration of basalt to hydrous minerals, e.g., chlorite, Livnat (1983), used δD and $\delta^{18}O$ to show that the basalts had undergone extensive isotopic exchange with low temperature meteoric waters prior to metamorphism and hydrothermal mineralization.

The Portage Lake volcanics are a classic example of low temperature mineral alteration, with the native copper deposits found associated with albitized wall rock and amygdule fillings of calcite, quartz, chlorite, epidote, prehnite, pumpellyite, laumontite, and agate. The mineralogy, textures, and absence of alteration mineralization indicate low pressure and temperatures (perhaps less than 100 °C). It is argued that metamorphism of lavas at depth altered basalt to pumpellyite, epidote, prehnite, chlorite and released copper (Guilbert & Park, 1986). Recently, hydrothermal mineral zones in the Portage Lake Volcanics have been established by oxygen and carbon isotope composition of calcite through inference (Bodden et al., 2022 and Fig. 7.12). In the Keweenaw Peninsula, four zones were identified: a Heulandite–Stilbite zone at temperatures of 100–150°; a Laumontite zone at 125–175 °C; a Pumpellyite zone at 175–250 °C; and an Epidote zone at 250–300 °C. A lower temperature zone at <125° consisting of thompsonite, scolecite, and smectite is found in the North Shore

Group volcanics in Minnesota. Since the mineral zone boundaries cross stratigraphic boundaries, the amygdule minerals formed after the occurrence of the basalt flows. The low temperature hydrothermal deposits of the native copper ores are found as a cement in conglomerate beds between the basalt flows, in fragmental brecciated and vesicular amygdaloidal flow top layers of individual flows, and in sparse fissures that cut both rock types. Thus, the ores are tabular and strata-bound and are in rocks that have high permeabilities (Guilbert et al., 1986).

Because of their higher stratigraphic position within the rift fill sequence, the degree of metamorphism and mineralization in the Lake Shore traps is much lower than in the Portage Lake volcanics. A zeolite facies is formed, as opposed to the prehnite-pumpellyite facies, that affected the flow top and deposited laumontite, analcite, calcite, smectite, chrysocolla, chlorite, datolite, and agate.

Further up-section, the Copper Harbor sediments represent an alluvial fan complex with basin-ward thickening of fluvial volcanogenic sediments that thicken basin-ward and distally fine up-section. Proximal and distal braided streams and sheet flood facies on coalesced alluvial fans and sand flats consist of lithic graywackes and siliclastic conglomerates that contain agate (Huber, 1975). The silicic volcanic clasts are derived from rhyolite bodies found on the south side of the Keweenaw Peninsula, indicating that there is a silica-rich component present in the Keweenawan series. The conglomerates are rounded cobble to boulder-sized volcanic clasts that are cemented with carbonate and iron oxide, and represent a fining upward, pro-grading, alluvial complex. The finer grained interbeds have cross-beds, current lineation, current ripples, foreset, and trough cross-beds, parting lineation, shallow medial fan lakes with algal stromatolites, and abandoned channels. The climate is arid, but with a high seasonal rainfall, and is positioned in an equatorial position. Laminated crypto-algal carbonate horizons, that are calcite-rich zones, represent the development of a vadose zone carbonate or paleo-caliche and white stromatolite (genus Colleria) horizons.

The extent of the Portage Lake Volcanics, the North Shore volcanics, the Lake Shore traps, and the Copper Harbor formation and amount of agatization is unknown as these rocks are under Lake Superior. It is thought, and probable, that the rift fissures, the source of the basalt flows, are in the center of the lake. Outcrops, then, are restricted to outcrops on Isle Royale, Michipicoten Island, the Keweenaw Peninsula, and North Shore of Lake Superior in Minnesota. Agate formed within the vesicles of the basalt flow tops (Fig. 7.13) and as clasts in the interbedded conglomerates of the Portage Lake basalt flows and Lake Shore Traps, indicating that agatization formed in the lavas before the conglomerates and sediments formed.

The silica source(s) and agate formation are problematic but are probably similar to other agate provinces with the dissolution of volcanic ash and tuffs on the surface and subsurface of the basalts by late-stage hydrothermal water and meteoric water. On Isle Royale, a volcanic tuff overlies the Greenstone flow of the Portage Lake Volcanic series and contains agates that have irregular shapes, similar to the thundereggs in the welded tuffs of the John Day formation in the Columbia River Plateau (Huber, 1975; Lane, 1911). The ash bed flows are porphyritic and scoriaceous with a notable clastic component. In some localities, the ash beds have pebbles and boulders of

a. Geologic Map Keweenaw Peninsula Native Copper District

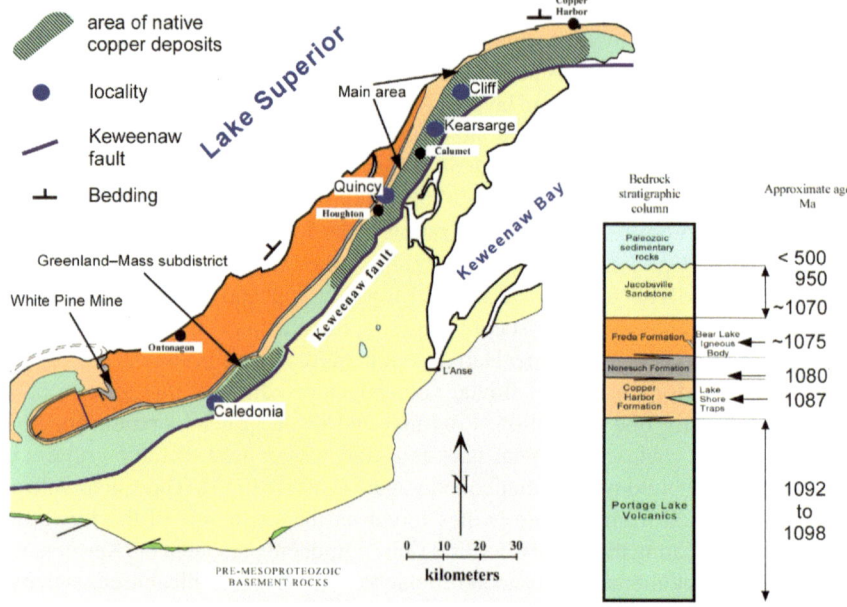

b. Main-Stage Hydrothermal Mineral Zones

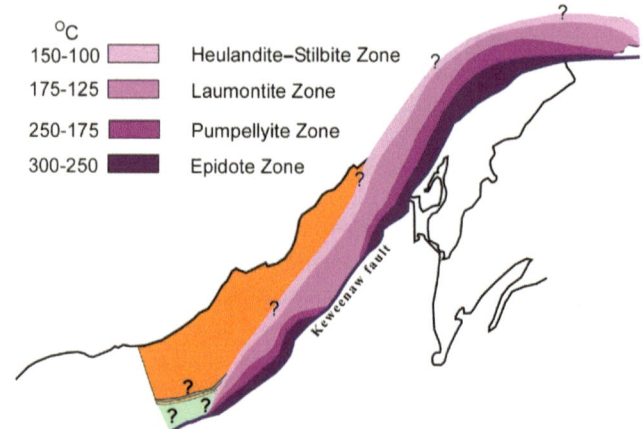

Fig. 7.12 Geologic map, stratigraphic column, and mineralization zones of the Keweenaw Peninsula (Bodden et al., 2022)

Fig. 7.13 Thin section of nascent amygdaloidal Lake Superior agate in basalt, 27 × 46 mm, ×2, polarized light

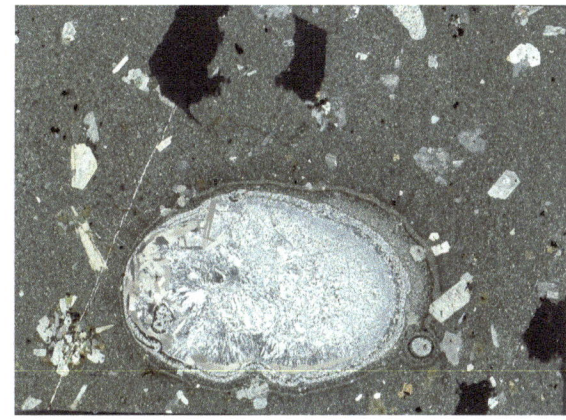

amygdaloid set in a sandy matrix. Broken pillow lava breccia (hyaloclastite) contains angular fragments of vesicular basalt that range in size from ash-sized to block size. Fragments have distinct rinds, while smaller fragments are finely fractured, having a perlitic-like texture. Mineralization of the ash bed contains calcite, quartz, epidote, pumpellyite, datolite, native copper and silver, laumontite, prehnite, adularia, analcime, apophyllite, faujasite, natrolite, and stilbite.

Whereas, agates are found in-situ on Isle Royale and Michipicoten Island, the majority of agates are found on the shores of Lake Superior, eroded, and washed up on shore (Fig. 7.14). The Lake Superior agate is a very distinctive agate with multiple fine bands of hematite (Fig. 7.15). The fineness of the banding indicates the great age of the agates. Aside from the Precambrian metamorphic agates of Russia and Australia, the Lake Superior agate is the oldest agate.

Fig. 7.14 Lake Shore trap basalt, Keweenaw Peninsula

Fig. 7.15 Lake Superior agate

7.4   Lake Superior Agates in Glacial Drift

Agates were redeposited in midcontinent glacial sediments and tills, from Minnesota to Kansas. The deposits of till, fluvial, and glacial outwash range from the source area of Lake Superior agates in the Keweenaw Peninsula and Thunder Bay, Ontario; to northeast and central Minnesota, northwest Wisconsin, northern Iowa, southeast Nebraska, and northeast Kansas (Fig. 7.16). Glacial ice as continental glaciation transported rock and deposited the rocks in glacial drift deposits.

In southeast Nebraska, Boellstorff (1980, 1988), through fission-track dating of interstratified volcanic ash beds that were deposited in former ponds and lakes, redefined the glacial stratigraphy of the area. Formerly, the major glaciations were termed the Nebraskan, Kansan, Illinoisan, and Wisconsin glaciations. In Nebraska, three major glaciations are found and termed the A, B, and C tills; the C till the oldest, and

Fig. 7.16 Sources of glacial erratics found in the midcontinent area (Lyle, 2009)

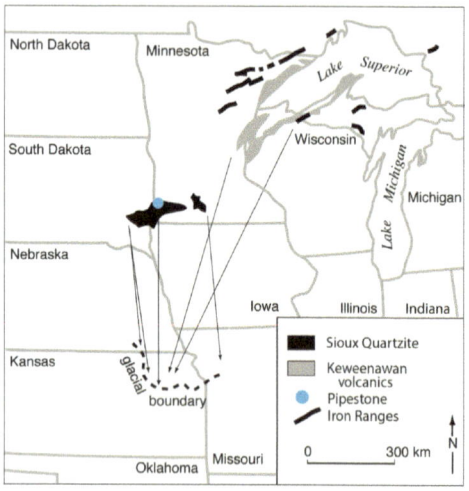

the A till the youngest. The fourth glaciation or Wisconsin glaciation is not present in Nebraska (Fig. 7.17).

The C till is considered to be the Nebraskan till, with two tills or glaciation pulses. Interglacial ashes between the C till and the B till are dated to 2–2.2 Ma. Agates are not in the Nebraskan or C tills, as the tills contain locally derived sedimentary rocks.

The "B" and "A" tills are considered to be the Kansan glaciation. "A" till was defined as having at least four advances and retreats of ice, with the oldest interglacial period between the B till and the A till dated from fission track dating of the Coleridge ash, to 1.9–1.5 Ma. Further up-section, the Pearlette ash and Hartford ash, was dated at 0.6–0.7 Ma. And further up-section yet, an ash was dated to 0.5–0.1 Ma. We are considered to be in the interglacial period that ended after the last glaciation pulse of

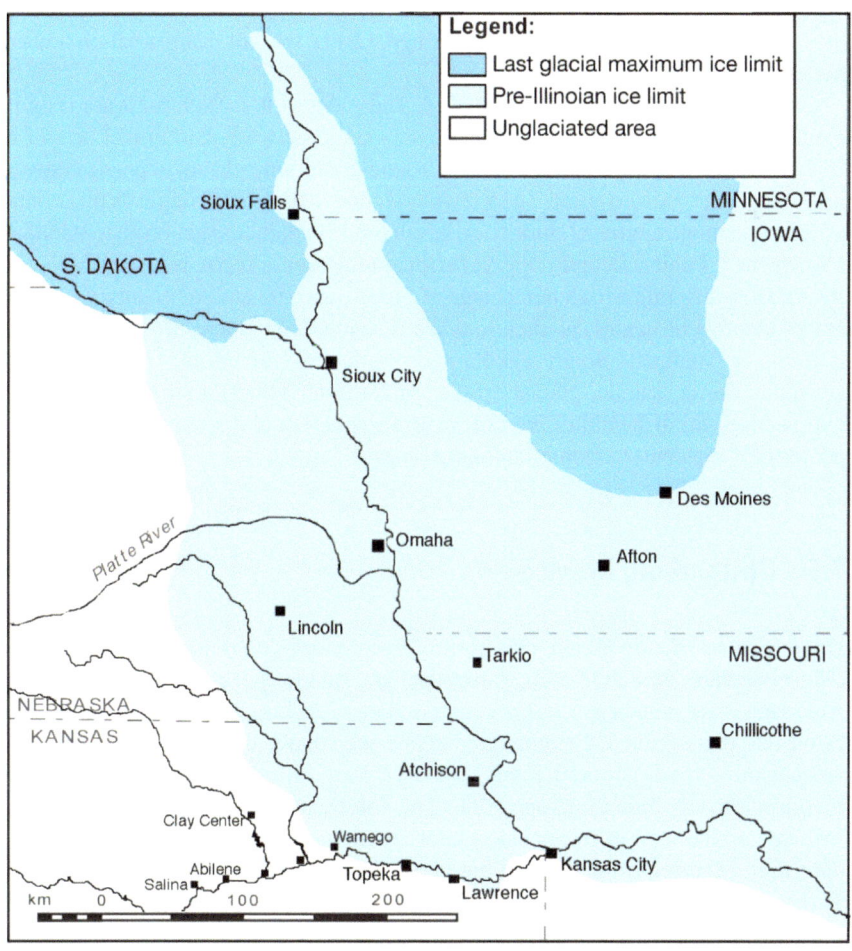

Fig. 7.17 Extent of glaciation in Eastern Nebraska Iowa, Missouri, and Kansas (Lyle, 2009)

Fig. 7.18 Glacial till over glacial outwash, Bennett, Nebraska (Courtesy of John Boellstorff)

the Wisconsin glaciation at 10,000 years ago. Or are we still in a glaciation advance period?

Agates are found in the B and A tills. These deposits in Nebraska are found in roadcuts, dirt non-graveled roads, sand pits, glacial outwash, and gravel bars of the Big and Little Nemaha River (Fig. 7.18). Agate-containing glacial deposits generally have a reddish oxidized color. The agates are found in the glacial deposits that contain a high percentage of green (chloritized basalt) and black rocks (gabbro?) that indicate a Keweenaw Peninsula and Lake Superior source areas. Topaz is also occasionally found. Deposits with a high percentage of Sioux quartzite and sedimentary rocks are not productive of agates, as they indicate a source area where the Sioux quartzite is found in Southeast South Dakota and Southwest Minnesota. Of interest, also, are found the occasional glacial erratics of catlinite. This is a silicious clay found interbedded within the Sioux quartzite in Minnesota. The clay was used by indigenous peoples for the construction of smoking pipes.

7.5   Chihuahua, Mexico

Igneous rocks of calc-alkalic composition were emplaced during Latest Eocene to Oligocene times with the subduction of the Farallon Plate (Pacific) beneath the North American Plate, which is rotating counter-clockwise (Fig. 7.19). The rocks are in an extension, plate-related silicic magmatic province that is contiguous with the silicic volcanism of the Basin and Range province to the north and east. The rocks are rhyodacite to rhyolitic in composition. The vulcanism is broadly contemporaneous with and gradational in composition between the calc-alkalic activity to the west in the Sierra Madre Occidental and the alkalic vulcanism to the east in Trans-Pecos Texas. The Sierra Madre Occidental Mountain range, the western boundary of the agatized areas, is noted for its vast volumes of ignimbrite flows, rhyolitic tephra, and related calderas that were emplaced in the Oligocene.

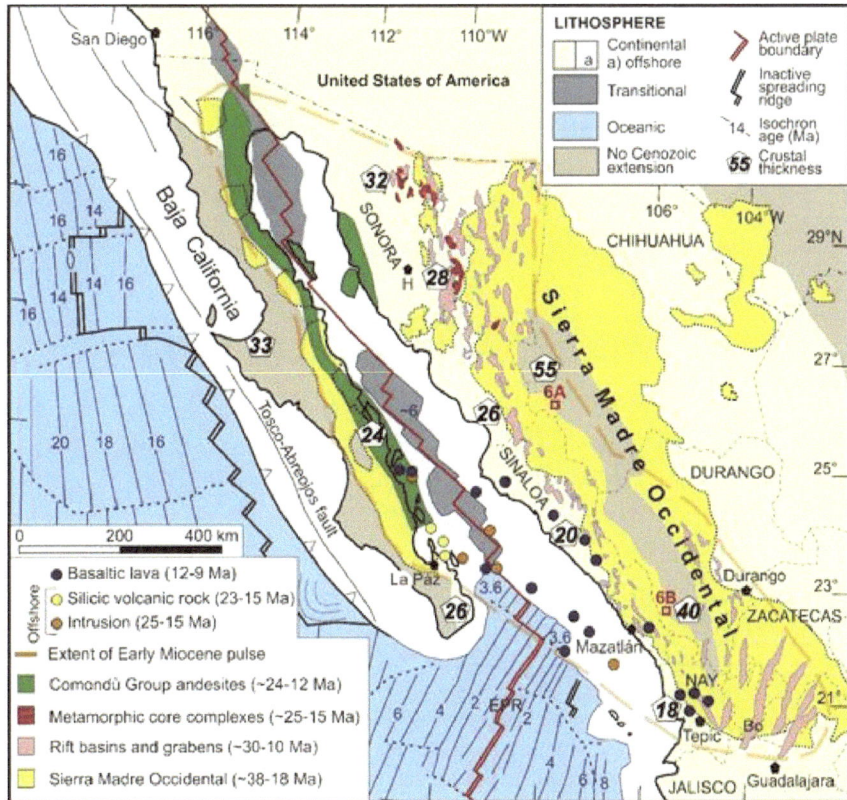

Fig. 7.19 Map of Mexico showing present-day configuration of plates, location of the Sierra Madre Occidental (SMO) and distribution of Oligocene to early Miocene volcanic rocks (Bryan & Ferrari, 2013)

The extensional tectonics in the pre-Basin and Range phase consisted of a change in tectonic strain from a NE to SW compression or extension during and just after Oligocene volcanism to an east–west extension and then to a northerly trending right lateral shear. This produced normal fault blocking and eastward sliding, and the folding and thrusting of Cretaceous sediments into the Chihuahua trough to the east. The regional structure consists of basin and range faults that trend north to northwest and cut all volcanic units. In response to the late Tertiary block faulting and subsequent erosion, the basin and range province was formed, consisting of coalescent desert basins with playa lakes, erosion of highlands producing a veneer of conglomerates, fanglomerates, colluvium, and bolson (basinal) fill of sands, clays, and gravels.

The vast majority of the agates are found in the basin and range block faulted Sierra del Gallego region bounded by the Sierra Madre Occidental mountains to the

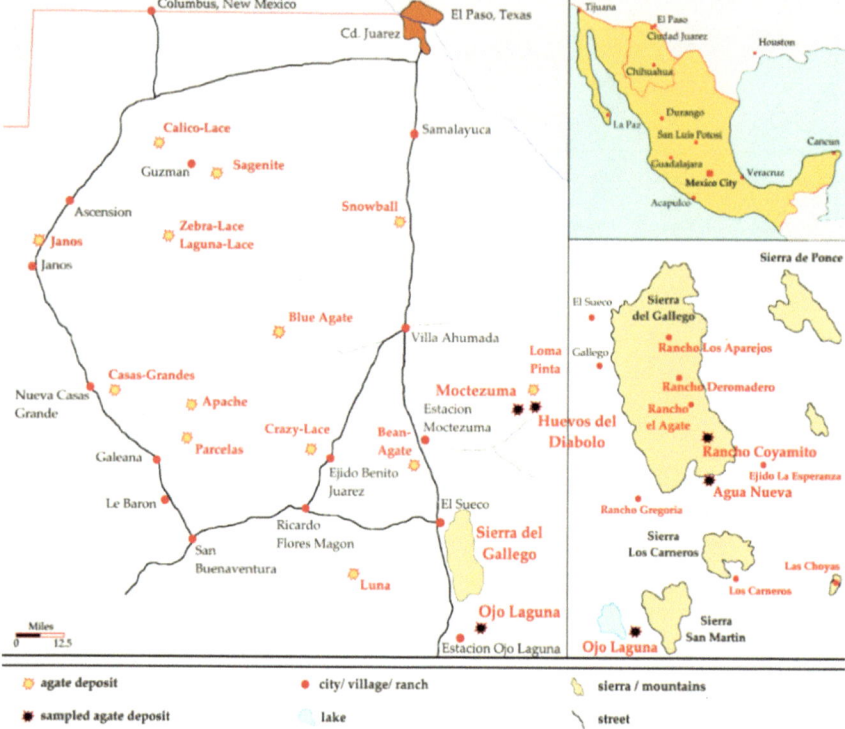

Fig. 7.20 Location map of Chihuahuan agates and the Sierra del Gallego Mountains agate area (Mrozik et al., 2023)

west and the folded Cretaceous carbonates to the east (Fig. 7.20). The area is 17 miles long and 2 miles wide.

The rocks in the region can be subdivided into four major types: (1) Tuffs and volcanoclastic sediments of the Libres Formation; (2) Voluminous intermediate-to-mafic lava flows of the Rancho el Agate andesite; (3) Rhyolitic units with lava flows, domes, and veins (Mesteño, Gallego, Agua Nueva, Carneros, and El Dos); (4) Basalt flows and flow tuffs of the Milagro Basalt (Fig. 7.21).

Zeroing in on the stratigraphy of the Sierra Gallego block; we find that it consists of the basal Liebres formation, which has a basal conglomerate member and an upper member, consisting of several thin ash fall and ash flow tuffs interbedded with tuffaceous sandstones. K–Ar dates are from 43.7 to 38.2 Ma in age; late Eocene in age (Keller, 1977). The upper Liebres tuff has had an intense silicification phase and a carbonate replacement phase.

The Rancho El Agate andesite formation has an age date of 38 Ma and overlies the Liebres formation. The andesite is the host rock for the agate that is found in the vesicles in the upper part of the lava flow. The andesite is an intermediate volcanic rock that consists of oligoclase, clinopyroxene, some quartz and opaque minerals.

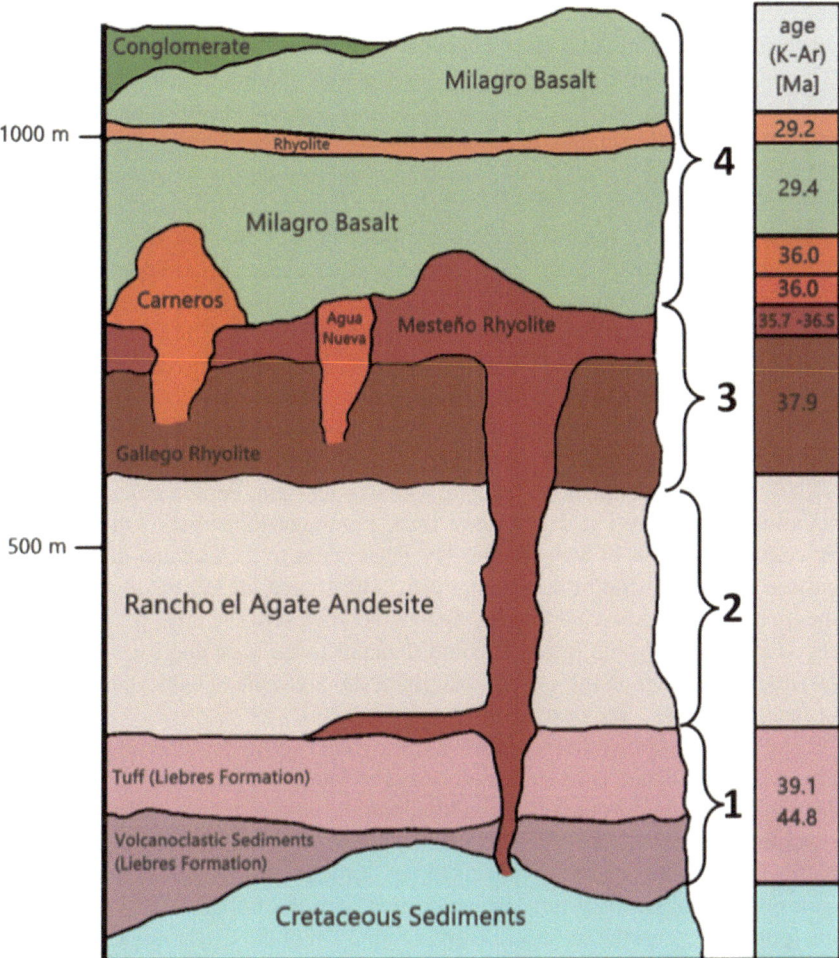

Fig. 7.21 Stratigraphic column of the Sierra del Gallego area, Mexico (Keller, 1977; Mrozik et al., 2023)

Twelve flows have been identified in the area; with individual flows 8 m thick on average. The uniform thickness of the andesite suggests a source of multiple vents with some pipe-like plugs and fissure vents. A source plug is found at Rancho El Agate, 10 km east of Gallego. The andesite forms rounded knobs on the flanks of mountains and a flaggy surface in valley floors, as at Hacienda Coyamito. The andesite is also present in the east in Sierra de Ponce and Cerra Ace Buche and to the west in Sierra de la Tinaja Lisa.

Except for the Ojo Laguna deposit in Sierra San Martin, south of Sierra del Gallego, all agate deposits are found in the same andesite lava flow as identified from major and trace element data (Mrozik et al., 2023). There are no flow textures in the andesite lava flow to indicate near surface flow from a different source. As usual, the agate amygdules are found in the upper part of the lava flow. Cathode Luminescence (CL) microscopy of Ojo Laguna agate indicated a 570 nm luminescence emission band indicating the presence of antimony at >300 ppm; also indicating that the agate formed during the alteration of the andesitic host rock as a later hydrothermal event (Mrozik et al., 2023). Antimony enrichment occurred during the formation process of the agate. Antimony is almost exclusively accumulated hydrothermally and is a chalcophile (affinity for sulfur) element and indicating that the Ojo Laguna agate formed with the involvement of hydrothermal solutions. In addition, the high Sb reading suggests a connection with the hydrothermal polymetallic ore deposits of Pb, Zn, Ag, Cu, Mn, and U deposits found regionally in the Chihuahua area.

The Gallego dacite, consisting of two flows, conformably overlies the andesite (Keller, 1977). The dacite consists of oligoclase, anorthoclase, and lesser amounts of clinopyroxene. Chill zones at the top and base of the lower flow consist of a black vitrophyre, that is highly devitrified to a fine-grained quartz and alkali feldspar, with the feldspars frequently altered to a chalky feldspar. Perlitic cracking and flow foliation are present. The dacite forms thick, columnar-jointed cliffs that overlie the layered round knobs of the andesite. The dacite thins to the east and is absent in the eastern part of the Sierra de Gallego area. The source area is unknown; however, the Sierra Chuchupate cauldron, by Yecora, southwest of the area, produced dacite, rhyodacite, and rhyolite flows. The unit thickens to the west and caps the Sierra de la Tinaja Lisa range in the western margin of the Sierra de la Gallego area. The age of the dacite is 37.1 Ma or early Oligocene.

The Mesteno domes and flows are rhyolites that overlie the andesite in the eastern area where the dacite is absent or eroded. The intrusives trend northerly to Ciudad Juarez and protrude from the bolson fill along the eastern margin of the area. These rhyolites form dikes and sills and exhibit a devitrification texture in the rhyolitic glass. The rhyolite occurs as a sill at the top of the Liebres formation and below the andesite. Minor ash flow tuffs are associated with the intrusives in the area. They are found 12 km northeast of Sueco and south 3 km of the Coyamito Ranch, and are generally less than 20 m thick, but do indicate a wide areal distribution. The age of the rhyolite is 35.7–34.5 Ma or Oligocene in age.

The Agua Nueva rhyolite is an intrusive that formed a 2 km wide circular feature that domed up the conglomerates and tuffs of the Liebres formation, the andesite, and dacite on the eastern side of the area near the Rancho Aqua Nueva. This rhyolite is 35.2 My or Oligocene in age. Barite-fluorite mineralization, along with some secondary silica mineralization in the fissure veins in the domed Liebres formation indicate some hydrothermal activity.

The Carneros rhyolite consists of a massive, 4 km plug with associated flows in the south–central portion of the area, in a valley north of Rancho Milagro. It is stratigraphically placed above the Mesteno rhyolite and has been emplaced at the same time as the Mesteno and Aqua Nueva rhyolites, around 35.1 Mya or Oligocene in age. The plug consists of steep columnar-jointed walls surrounding a central interior basin that was differentially eroded.

There are associated flows to the north, east, and south. The flows have a white ashy, agglomeratic base. The flows thicken to the east, with blocks of dacite and Mesteno rhyolite in the flows. A large flow trends south–southwest from the plug.

The Milagro basalt consists of 25 flows with vesicular tops with calcite and zeolitic fillings. It is best exposed at Rancho Milagro, 30 km se of Sueco. A 10 m thick, unaltered rhyolitic ash flow tuff is on top of the basalt. It is bimodal in composition, consisting of a basalt and ash flow tuff. The basalts form a wide plateau dissected by basin and range faults. A shield volcano 12 km west of Sueco could be a possible source of the basalt. The basalt is 28.7 Ma in age and the rhyolite tuff is 28.5 Ma in age or Oligocene in age and are the last volcanic flows in the area.

A fanglomerate, with all rock units found as clasts in the fanglomerate, is the alluvium/bolson fill in the area. It consists of unconsolidated sediments of clay, sand, and gravel developed during local up-lift and fault growth that occurred as late as 5 Mya.

7.5.1 Agates from the Sierra Gallego Area

Rancho Gallego agates are from the rounded eastern slopes of Cerro del Gallego (10 km east of Highway 45) (Fig. 7.20). They are found in the altered El Rancho Agate andesite and have a characteristic red and gray banding.

The Coyamito is the earliest agate found in the area, dating back to 1946. It has purple, rose, yellow, white, red, mustard, and dark lavender banding (Fig. 7.22). The agate has frequent agate pseudomorphs after hexagonal, twinned aragonite and has a limonite exterior.

Fig. 7.22 Coyamito agate

Fig. 7.23 Laguna agate

The Rancho Aqua Nueva agate area borders on the south side of the Rancho Coyamito. The Aqua Nueva agate is mined from a tunnel driven into the andesite that makes up the northern flank of Cerro de la Aguja. The agate is nodular and is also found in veins. The agate has a characteristic lavender, purple, gold, pink, and yellow banding.

The Aparejos agate is found at the Rancho Los Aparejos from hand-dug pits, dug into a very dense El Rancho Agate andesite. It has characteristic purple and yellow bands.

The Gregoria agate is found at the Cerro El Ahijadero from hand dug, very dense Rancho El Agate andesite.

Laguna agate (Fig. 7.23) is from Rancho Borunda at Ojo Laguna, 20 km south of Rancho Agua Nueva, adjacent to the eastern shore of Laguna Encinillas. It is mined from three pits in the colluvium that is derived from the andesite, that makes up the lower half of the Sierra San Martin, just east of Laguna Encinillas. The agate consists of fortification banding in shades of red, blue, green, and yellow.

7.5.2 Las Choyas "Coconut" Geodes

The geodes are found 35 km ENE of Laguna Encinillas. They occur in the basal vitrophyre of the Liebres formation that has been altered to bentonite. It is known as the "caliche" layer. The age of the Liebres formation is 43.7 Ma in age or Late Eocene. The geodes occur in irregular to spherical lithophysal cavity fillings, similar in occurrence as the thundereggs in the John Day formation of the Columbia plateau. Water vapor collects in the spherulites of alkali feldspar and cristobalite, that resulted from devitrification of the ash flow tuff. Circulation of vadose groundwater leached silica from the vitrophyre and altered the vitrophyre to bentonite. A chalcedony lining in the cavity grades into mega crystalline quartz, that grades in color from clear to smoky to amethyst. Late-stage minerals occur in the geode center that consist of carbonates, manganese oxides, calcite, zeolites, iron oxides, and hydroxides. Silt deposition, with various sedimentary structures such as graded bedding, cross-bedding, and slump structures, indicates probable meteoric, vadose groundwater circulation. Oxygen isotope analysis provided the temperature of geode formation with temperatures ranging from 48.5 to 78.2 °C (Keller, 1977).

7.5.3 Conclusions

An unconformity between the Mesteno, Agua Nueva, and Carneros rhyolite volcanism phase and the Gallego dacite indicates an erosional and weathering interval. The unconformity extends to the Rancho El Agate andesite where the Gallego dacite is absent. The Gallego dacite thins to the east in the area and is absent along the eastern margin of the area and thickens to the west where it caps the Sierra de la Tinaja Lisa. The Mesteno domes in the area are associated with extensive flow-foliated lava flows of the same composition that everywhere overlie the Gallego dacite, except where the latter is absent, as along the eastern margin of the area where the Mesteno lavas overlie the Rancho El Agate andesite.

The andesite host rock has been weathered to iron oxides and manganese oxides that form the colors and banding in the agates. The source of silica for the agates probably came from an eroded ash fall or ash-flow tuff in the Gallegos area. The Mesteno, Carneros, and Agua Nueva rhyolite events suggest a rhyolitic/ash fall/ ash flow phase occurring 35.7–34.5 Mya. Possible sources of this rhyolitic phase volcanism are the Sierra Chuchupate cauldron near Yecora, southwest of the area, that spewed acidic rocks as air fall ash, ash flow tuffs, dacites, rhyodacites, and rhyolites; and the Sierra Pastoias area, south of Ciudad, Chihuahua, that spewed 35 My ash flow tuffs.

Six miles west of Ojo Caliente, near the Luna agate locality, is a road cut with an andesitic conglomerate overlying bentonitic ash bed with relict pumice texture; perhaps an erosional remnant of a once widespread deposit that overlaid the Rancho El Agate formation (West Texas Geological Society, 1974 Field trip Guide). The Rancho El Agate andesite extent is unknown. It is uniformly thick throughout the area, with thick exposures present to the east in Sierra de Ponce, Cerra Ace Buche, and to the west in Sierra de la Tinaja Lisa. The andesite does not seem too thin to the north or south, suggesting a large areal extent that incorporates other agate localities, such as the Luna agate, Moctezuma agate, the Loma Pinta agate, Apache, and Parcelas agate.

The conglomeratic member of the andesite suggests an erosional period with braided streams, channel scouring, water reworking of sediments, and silica leaching of air fall ash. The bentonitic beds indicate leaching of silica from volcanic ash. Oxygen isotope data indicate paleotemperatures that range from 30 to 68 °C (Keller, 1977), suggesting meteoric and vadose water components. Except for the Ojo Laguna agate with the high Sb content and the small area of barite–fluorite mineralization that occurred in the Liebres tuff dome at the Rancho Agua Nueva, the mineralogy, textures, and the absence of hydrothermal alteration minerals, suggest low temperature mineral formation from mixed late-stage near surface hydrothermal waters and meteoric and vadose waters. Trace element data from agate samples of Al, Fe, K, Na, Ca, and U indicate formation of the agate from near surface weathering solutions (Mrozik et al., 2023). Also, as thorium is immobile under weathering conditions, the Thorium/Uranium ratio argues for the mobilization of silica under weathering conditions (Mrozik et al., 2023). The age of the Mesteno rhyolite is 35.7 Ma, and the age of the Gallegos dacite is 37.1 Ma, indicating an erosional and agatization period of 1.4 Ma.

7.6 Brazil–Parana Basin Namibia–Etendeka Plateau Flood Basalt Provinces

The Parana and Etendeka basins formed 129–132 Mya in the early Cretaceous. The separation of South America from Africa and the opening up of the South Atlantic Ocean occurred approximately 120 Mya. Heat from the Walvis hot spot under the Parana region of eastern South America and present-day Namibia in Southwest Africa (continents were not yet separated) caused doming of the crust (Figs. 7.24 and 7.30). Consequently, deep fractures formed through which lava poured into the surface. The continental flood basalts are approximately 1,800 m thick and 1 million km^2 in area and are composed of tholeiitic basalt, dacites, andesites, and latites (Peate, 1997).

Rhyolitic eruptions, erupted as high temperature ash flows, accompanied the final magmatic phase with the formation of several alkali complexes and associated collapsed calderas on the margins of the Parana and Etendeka basins (Peate, 1997).

The flood basalts form the Serra Geral Formation in Brazil, forming a plateau with no topographic relief, and little to no suitable sections available for flow-by-flow description. While the multiple lava flows appear to be homogenous, in reality, they are really heterogeneous within a single lava flow, exhibiting features such as vesicular zones, vertical and horizontal jointing, vitreous zones, tabular (columnar jointing) classic flows, hyaloclastite zones, and bole or weathered surfaces. The lavas erupted subaerially onto the Jurassic Botucatu formation over virtually the whole basin (the Parana and Etendeka basins). The Botucatu formation consists of aeolian sandstones that persist as intercalations and xenoliths up to 160 m thick, within the Parana and Etendeka lava sequences. The climate was arid and desert-like as suggested by the presence of aeolian sands, lack of paleosol development, and the absence of entablature joint patterns in the basalt lava flows, indicating the absence of percolating water during cooling of the lava flows (Strieder, 2006).

The agatized area is in the Salto do Jacui mining area in the Rio Grande do Sul State and is ~160 km^2 in area. Seven volcanic units are present in the Salto

Fig. 7.24 Parana and Etendeka basin volcanics and plate separation (after Machedo et al., 2018)

do Jacui mining district where the agates are mined (Figs. 7.25 and 7.26); from oldest to youngest: a Lower Dacite unit, a Lower Basalt flow, a Lower semi-glassy vesicular, amygdaloidal dacite flow, a vesicular amygdaloidal basalt flow, a vesicular dacite flow, an altered volcanic glass zone, an Upper semi-glassy dacite flow, and an Upper Dacite flow (Strieder & Heemann, 2006). The agate is found mainly in the vesicular, amygdaloidal basalt flow, generally 4–6 m thick. This agatized basalt unit also contains Botucatu formation intercalations that formed during moments of magmatic acquiescence and xenoliths that formed when the basalt flows initially flowed over the desert dunes of the Botucatu formation. Agate is also found in the lower semi-glassy dacite unit but is not as abundant as in the basalt flow.

According to Strieder and Heemann (2006), the lower unit, where agate mineralization occurs, is a weathered lava flow, with three different zones. The lower zone is more mineralized with agate than the middle zone, and is a basaltic/andesitic rock

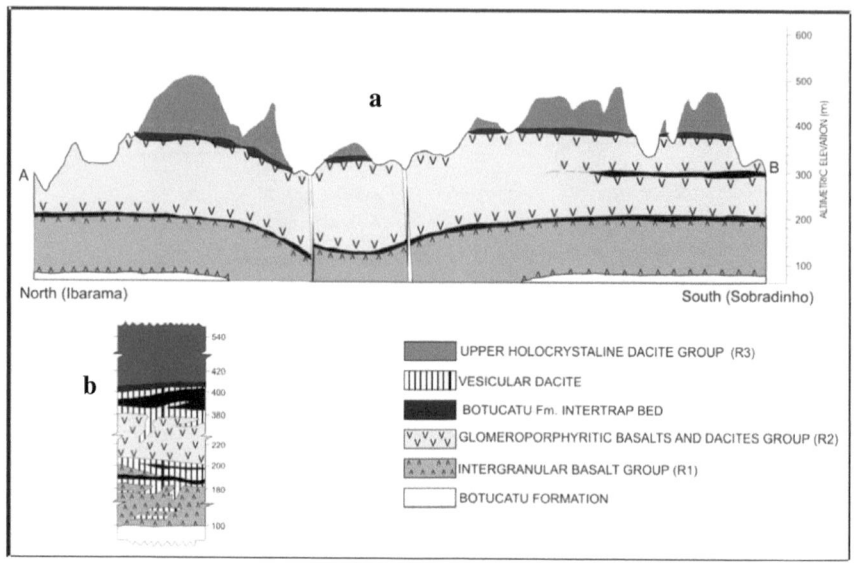

Fig. 7.25 A N-S cross-section in the Salto do Jacuí region from near Sobradinho to Ibarama. **B** Columnar section (after Strieder & Heemann, 2006)

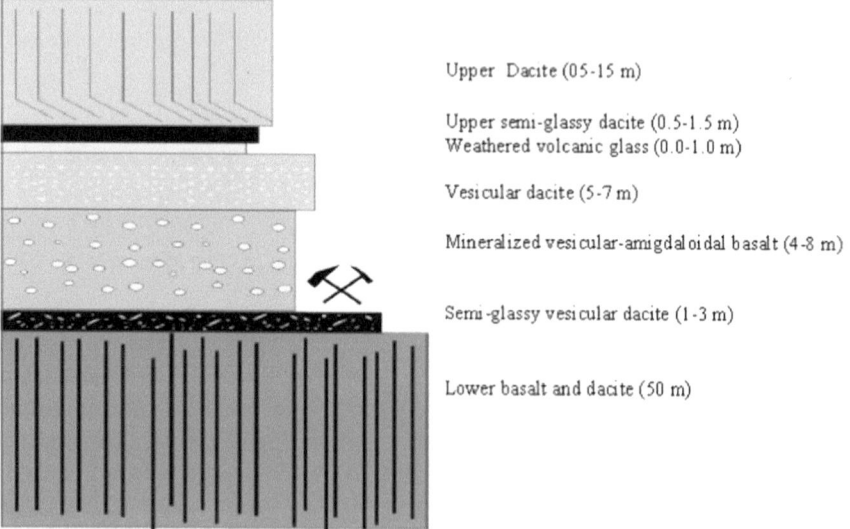

Fig. 7.26 Salto do Jacuí mining district columnar section, showing agate interval (Streider & Heemann, 2006)

Fig. 7.27 Typical Brazilian
agate with broad diffuse
banding

of gray to brownish gray colors, with a thickness of 2–4 m. There are remnants of
unweathered, black glassy rock. The middle zone is an altered, aphanitic, reddish
brown to yellowish brown dacitic rock that is 2–3 m in thickness. The upper zone
is 0.5–2 m in thickness and is a yellowish to grayish yellow color. It is a strongly
weathered rock, which originally may have been a glassy volcanic ash. The upper
unit is 2–4 m thick and is an aphanitic, gray to brownish-gray, weakly altered dacitic
lava flow, with a regular horizontal fracture pattern.

Agates are found up to 0.8 m in diameter in the vesicular basalt and the overlying vesicular dacite. Brazilian agates characteristically have broad diffuse banding
(Fig. 7.27). The agatization sequence proceeds from chalcedony to quartz crystals.
Associated minerals found within the agate are calcite, with manganese and iron
oxides. The oxides can occur as moss or dendrites.

The intercalated and the xenolithic, epiclastic sandstone, and volcaniclastic sandstones and conglomerates of the Botucatu formation show fluid flow textures and a
secondary and alteration mineralogy from late magmatic fluids (Strieder & Heeman,
2006). They inferred that late magmatic fluids incorporated sand from the Botucatu
xenoliths and intercalated lenses, melted the clastics, and provided the silica for
the agates. As some agates show some banding, consisting of alternating layers of
sandstone and/or glassy rock and agate, these features suggest a link between the
sandstone and agates.

The secondary mineralization was evaluated and sequenced in the equivalent
Botucatu African deposits by Rios (2023) (Fig. 7.28). This genetic sequence is
identical to the agatization mineral sequence. We have illite and smectite clays,
celadonite, opal, chalcedony, and zeolites (apophyllite, stilbite, mordenite, analcime
and chabazite) formed. Interestingly, there are no agates found within the sandstones

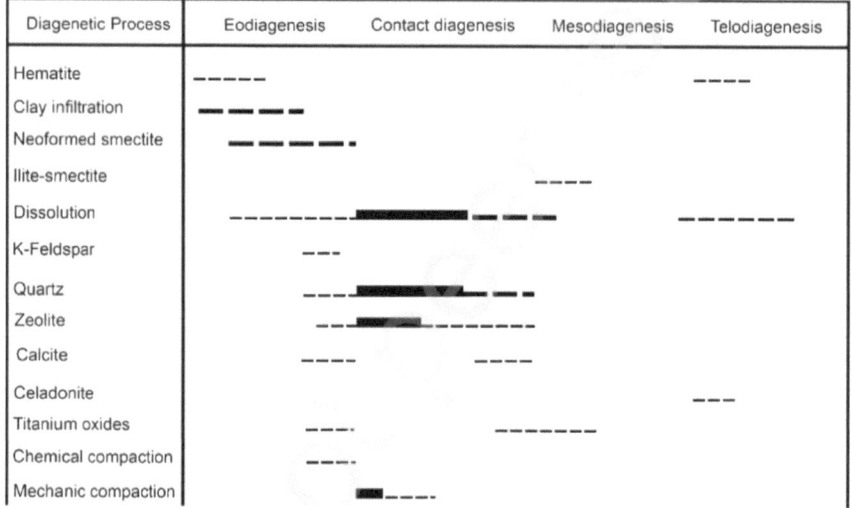

Diagenetic Process	Eodiagenesis	Contact diagenesis	Mesodiagenesis	Telodiagenesis
Hematite	------			----
Clay infiltration	------			
Neoformed smectite	-------			
Ilite-smectite		----		
Dissolution	------------------	--- --	-------	
K-Feldspar	---			
Quartz	--- ▼--- --			
Zeolite	--- ----------			
Calcite	----	----		
Celadonite				---
Titanium oxides	----·	-------		
Chemical compaction	----·			
Mechanic compaction	----·			

Fig. 7.28 Paragenetic mineral sequence for volcaniclastic and epiclastic sandstone in Botucatu Formation, Brazil (Rios et al., 2023)

and conglomerates, but a "proto" or semi-agate is found. This secondary mineralization sequence, present in the Botucatu formation, was deposited by late magmatic waters and vapors. However, considering the huge volume of agate deposited in the basalts, this method would not produce enough silica to produce the vast quantity of agate found here (Rios, 2023). It is unlikely that the silica originated from late magmatic waters and vapors, but originated from ash deposits that were altered by heated meteoric waters during an erosional and weathering cycle.

With the rifting and sea floor spreading of Western Gondwanaland 120 Ma in the Cretaceous period, the African counterpart to the Brazilian Serra Geral formation is found on the Skeleton Coast of Namibia on the Etendeka plateau (Fig. 7.24). The Karoo volcanic sequence consists of latite and quartz latite flows capped rhyolitic eruptions in a final magmatic phase (Peate, 1997, and Figs. 7.29 and 7.30). The volcanic pile rests on the Karoo sedimentary sequence that is identical to the Brazilian Botucatu Formation. The sediments thicken to the coast, are synsedimentary with the inland sediments and have parallel faulting, with the faults active during and after the building of the Etendekan volcanic pile. Ephemeral lakes were formed during the outflow of the volcanic flows. The agates are found in the latite flows.

Harris (1989), calculated $\delta^{18}O$ values from Etendeka agates, that range from $20.4^{0/00}$ to $28.9^{0/00}$. According to Harris, these values correspond to temperature values of 26–169 °C, averaging ~120 °C, indicating mixed heated meteoric and late phase magmatic waters. Brecciation fragments found in some agates were found to be evidence of boiling water textures and mixing of silica-rich meteoric and late phase magma waters and vapors. The depositing fluids circulated at shallow levels in a cooling lava pile with prolonged boiling and loss of vapor, rather than with the

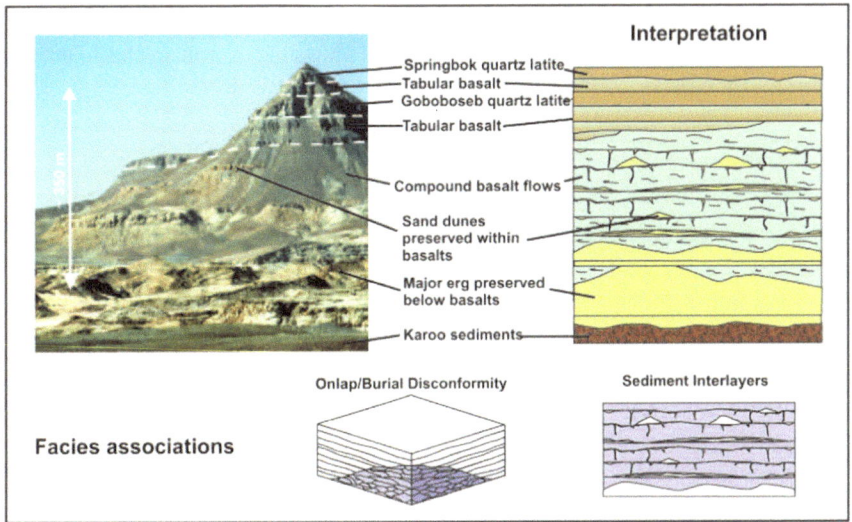

Fig. 7.29 Etendeka stratigraphic column (Nelson et al., 2009)

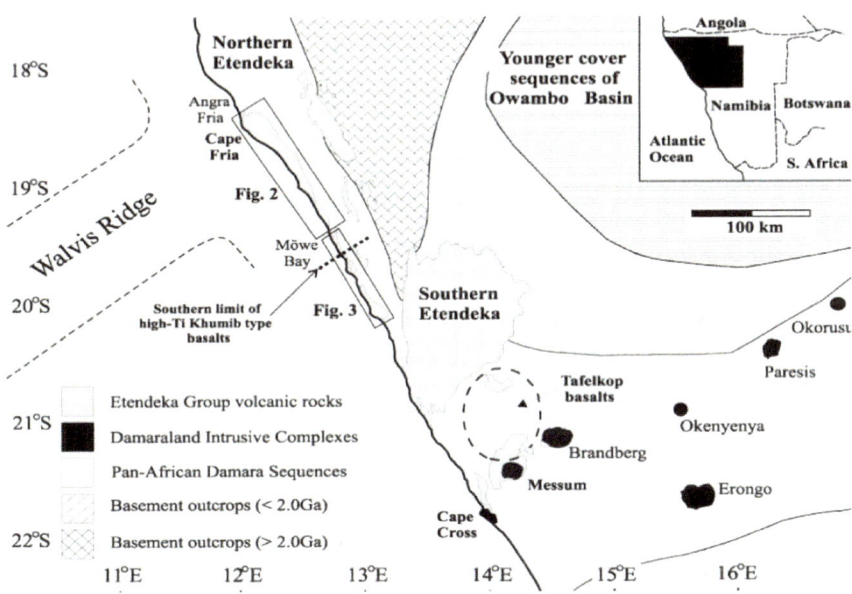

Fig. 7.30 Geologic map of Etendeka Flood Basalt Province, Namibia, Africa (Ewart et al., 2004)

exchange of fluids with the host volcanic rock. With changes in the pressure and temperature due to changes in the hydrothermal plumbing or from cycles of eruption followed by erosion above the agate zone, the silica was deposited and the agates formed.

7.6.1 Brazil Amethyst Geodes

The most important amethyst mining region is in the northern part of the Rio do Sul State and is the Ametista do Sul Mining District. The deposits occur in an area of about 300 km^2, which encloses more than 350 mines. In this mineral district, four amethyst-bearing lava flows in the Serra Geral Formation were identified (Juchem, 2014).

The general stratigraphic sequence consists of a basal 5–10 m of strongly weathered vesicular basalt with enclosed Botucatu sandstone blocks and xenoliths at the base, overlain by 0.3–1.5 m of highly fractured aphanitic basalt; followed by 20–30 m of massive basalt of which, in the top 2–4 m, occurs the major geode horizon. Overlying the main geode-bearing layer are 5–7 m of vesicular rhyodacite, 5–8 m of basalt; and 2–3 m of a vesicular glass-rich dacite, all of which contain sporadically occurring geodes (Gilg et al., 2002).

A more detailed stratigraphic column (Fig. 7.31) shows two lava flows. The lower basalt shows four zones consisting of a thin lower vesicular zone (~0.50 m thick), that transitions to 10 m thick, massive gray to greenish-gray basalt, with a vertical and horizontal jointing. Overlying the massive basalt zone is the major geode horizon. Above this horizon, there is an irregularly jointed basalt, 0.5–1.0 m thick, followed by an altered, horizontally jointed basalt level that is 0.5–1.0 m in thickness with a horizontal jointing pattern. The upper zone of the lower basalt flow is 1–5 m in thickness and is an altered light-gray to dark-gray vesicular basalt. Between the two flows are layers of Botucatu sandstone and/or breccia that are a few centimeters to 2 m thick. The breccia is composed of irregular clasts of sandstone or basalt, cemented by lava, usually mixed with sand and microcrystalline quartz (Juchem, 2014).

The geodes range in size from a few cm to 6 m. The geodes consist of a green botryoidal celadonite rim, then agate, rarely as stalactites (Fig. 7.32). Brecciated textures are possible, consisting of detached basalt and celadonite fragments from the wall of the geode and then cemented by agate. Ductile deformation zones (escape or infiltration tubes) are present followed by euhedral quartz, that is clear and amethystine. Rarely are there more than two or three cycles of agate deposition. Calcite and rarely gypsum formed throughout the paragenetic sequence. The celadonite forms the outer shell of the geodes and is formed from the alteration of basalt. Calcite, lepidocrocite, and goethite needles are found in the amethyst crystals. Very rarely barite crystals or pseudomorphs of quartz after anhydrite are found in the central cavity (Gilg et al., 2002).

The host rock is a tholeiitic basalt with a high TiO$_2$ content, found in the basal portion of the volcanic succession. The basalt consists of hypocrystalline, aphyric to

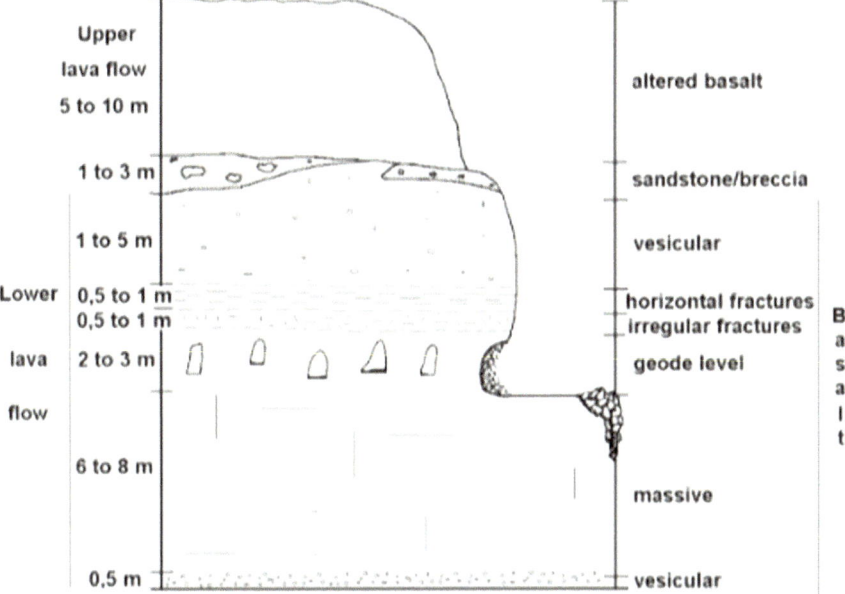

Fig. 7.31   Amethyst geode stratigraphy (Juchem, 2014)

Fig. 7.32 Brazilian
amethyst geode

subaphyric plagioclase, augitic clinopyroxene, rare altered olivine in a partly glassy, partly crystallized, silicic groundmass that exhibits local argillic alteration. Oxygen isotope temperatures from two phases in a quartz crystal geode, indicated waters at temperatures of 95–98 °C. Calcite samples ranged from −18.7 to −2.9 0/00 or <100 °C. According to Gilg, the source of the silica is thought to be the interstitial glass of the host basalts, leached by gas-poor aqueous solutions of meteoric origin, ascending from the locally artesian Botucatu aquifer located in the footwall of the volcanic sequence.

Geode formation was a two-stage process involving early magmatic proto-geode cavity formation and a later, low temperature infilling of the cavity. An aqueous or carbonic vapor phase that is exsolved from the basaltic lavas is the immiscible fluid responsible for the cavity formation; with a massive bubble coalescence that is responsible for the huge cavities of the geodes. The cap-like features and basal dimple indentation indicated ascent of fluids with a lower density and viscosity than the surrounding medium (Gilg et al., 2002).

Vasconcelos (1998) provided ^{40}Ar ^{39}Ar age dates of the geode celadonite. They are 65–70, 80–90, and 110–113 Ma and indicate a long period of celadonite formation. It took 40–60 Ma after the basalt extrusion to form the geodes.

According to Gilg, fluid contributions from the Botucatu sandstone aquifer are not excluded. The geode in-filling process probably started with the onset of sustained freshwater recharge to the Guarani (Mercosol) aquifer that is a Jurassic to Early Cretaceous eolian sandstone. The process likely started in mid-Cretaceous times and ended ~40–60 Mya after the basalt extrusion, with the circulation of artesian, meteoric water. The initial freshwater recharge from the southeastern margin of the basin and establishment of the present-day flow net and hydraulic gradients started 110–90 Mya as a consequence of initial uplift of the coastal Serra do Mar Mountains, that is related to the opening of the southern Atlantic Ocean.

7.6.2   Petrified Forest

Petrified wood is formed from dissolved volcanic ash and is an excellent indicator of the presence and dissolution of volcanic ash, under surface conditions, and with circulating meteoric waters; thereby, providing a large silica source (see Arizona Petrified Forest section). A petrified forest is located in the central region of Rio Grande do Sul State, in the Santa Maria region, approximately 100 km south of the Salto do Jacui agate area. The petrified wood is in the Triassic sedimentary rock formations of the Marta, Santa Maria, and Caturrita Formations. The forest consists of mesophytic flora of the Triassic period. The petrified forest suggests that volcanic ash deposits were in the region and could be the silica source of the Salto do Jacui agates.

7.7 Watchung Basalt, New Jersey Agates

Late Triassic and Early Jurassic Newark Supergroup basins formed during the rifting and opening of the Atlantic Ocean (Fig. 7.33). Flood basalts associated with the break-up and rifting of the Pangea supercontinent occurred with three Watchung basalt lavas extruded in rapid succession in the northern New Jersey area during the Hettangian age of the early Jurassic period (Fig. 7.34). The basalt flows alternate with Jurassic sedimentary rocks. Subsequent folding and faulting tilted the rocks to the west with dips 5–25° to the west, forming the Watchung syncline (inset cross-section Fig. 7.34). Fissures, such as lava filled feeder dikes, from which the lavas flowed into the basin, have not been found. As this area is highly urbanized, the fissures are probably underneath a parking lot somewhere.

Agates are found in the Orange Mountain basalt. It is a characteristically fine-grained, pillowed, pahoehoe-like, vesicular, amygdaloidal, tholeiitic, olivine-poor basalt with at least two flows identifiable (Olsen, 1980). Where not vesicular and pillowed, the flows have a columnar entablature. The upper and lower flows of the Orange Mountain basalt are separated by a thin volcaniclastic bed, generally <1–4 m thick with numerous beds of ripple-bedded mud-cracked siltstone. The lava flows are a quartz normative basalt composed mainly of plagioclase and augite with minor amounts of orthopyroxene and altered olivine. The basalts are chemically similar to Lesotho basalts of the early Jurassic Karoo volcanic province, South Africa.

Low temperature magmatic fluids containing copper filled the vesicles of the basalt flows. Later secondary mineralization fluids deposited zeolites (Puffer et al., 1992). Zeolite minerals found include stilbite, heulandite, chabazite, pectolite, analcime, datolite, and prehnite. A later erosional and weathering cycle precipitated the sulfate minerals, gypsum and glauberite, and the carbonate, calcite, in the cavities of the basalt. These minerals are indicative of an arid, evaporative environment, such as playa lakes. Clays are present as later alteration minerals.

The Passaic formation underlies the Orange Mountain basalt (Olsen, 1980 and Fig. 7.34). It is predominantly red clastics; red siltstones, sandstones, and conglomerates. Above the Orange Mountain basalt is the Feltville formation that overlies the eroded surface of the basalt. It is early Jurassic in age. It consists of a red siltstone and a buff, gray, and white feldspathic sandstone, with fining upwards sequences. The lower half of the contains a black to white laminated calcarenite limestone and a graded siltstone bed containing abundant fossil fish, sandwiched between two beds of gray, small to large scale cross-bedded siltstone. Further up-section is a conglomerate bed, located 15 m beneath the overlying Preakness basalt, that consists of vesicular basaltic clasts, phyllite and limestone pebbles and cobbles. The basalt clasts can be quite numerous, consisting up to 30% of the conglomerate (Olsen, No agate has been found in the Feltville conglomerate.

No silica source is discernable in this area. As noted by Darton (1891), there is a general absence of volcanic ash and tuff. The erosional surface of the top flow of the Orange Mountain basalt indicates that whatever originally overlaid the basalt— probably an ash or tuff unit, is gone. The feldspathic sandstone of the underlying

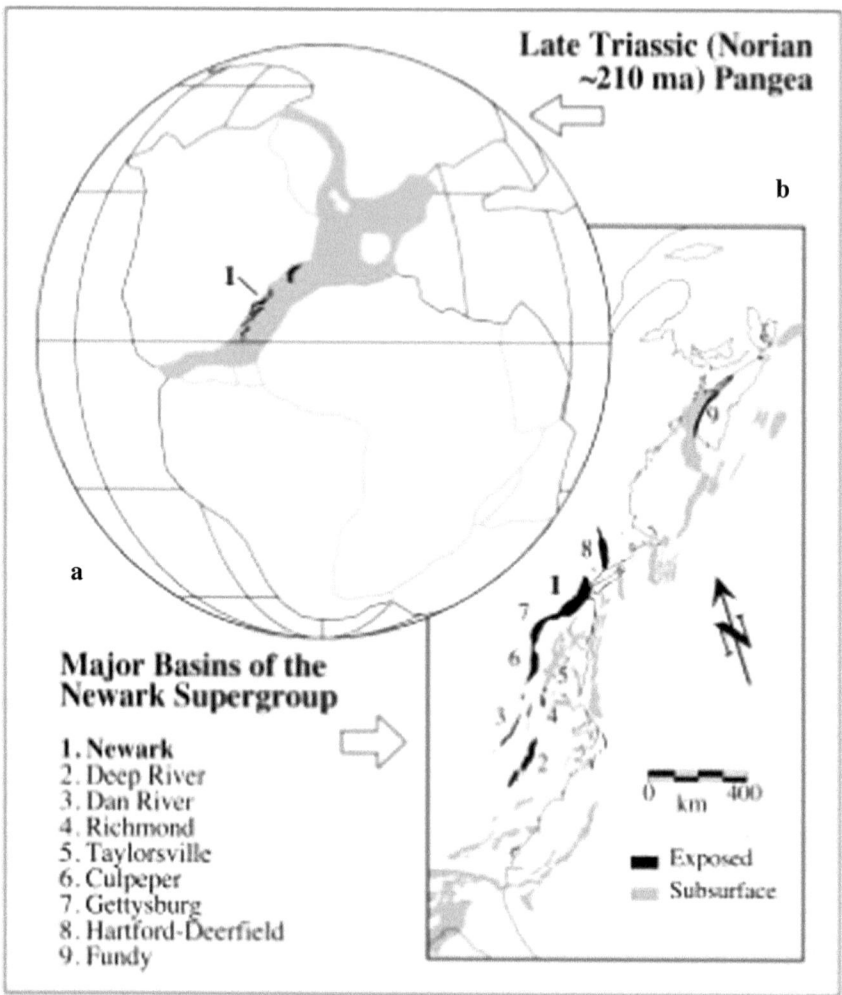

Fig. 7.33 Triassic–Jurassic Newark Supergroup Basins East Coast of North America (Olsen, 1999)

Passaic Formation contains abundant feldspars and could be a silica source locally. However, no altered feldspathic sandstone has been located. Agates are found in a localized area, namely the Prospect Park Quarry and the Upper New Street Quarry. The agates are a pink to red banded agate, with the red coloring caused by hematite, the weathered oxide mineral found in vesicles of the basalt (Fig. 7.35).

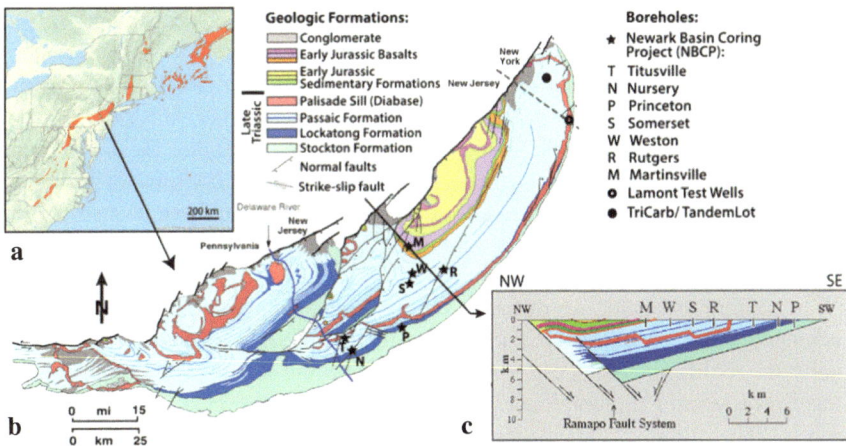

Fig. 7.34 Newark basin location, geological map, and cross-section (after Zakharova & Goldberg, 2021)

Fig. 7.35 A characteristic pinkish, reddish Orange Mountain basalt agate

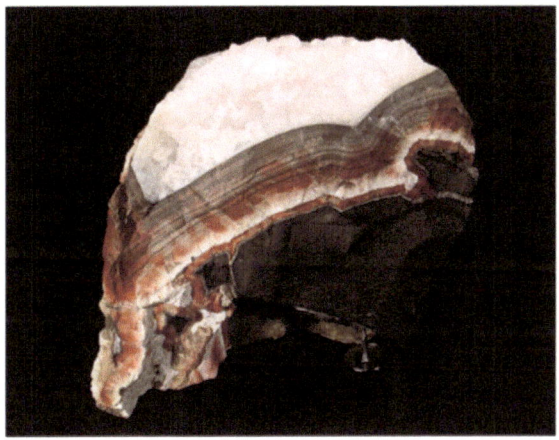

7.8 Nova Scotia

Agates also occur in the same geologic province as the New Jersey Watchung agates, occurring in the earliest Jurassic (Hettangian) North Mountain basalts of the Newark super group in the Bay of Fundy, Nova Scotia (Figs. 7.33 and 7.34). In the Middle Triassic, the Pangea supercontinent broke up, with continental rifting producing volcanism in grabens and half grabens. Agates, celadonite, and zeolites are found within vugs, amygdules, and fractures of the basalt.

The Minas Basin on the north side of the Bay of Fundy is situated within a half-graben at the eastern end of the bay, bounded by a major fault of the Cobequid-Chedabucto fault system, that with a strong left-lateral movement coincided with basalt extrusion (Figs. 7.36 and 7.37). The basalts are considered to be tholeiitic flood basalts that erupted from fissures and are very similar to the New Jersey Watchung basalts. The basalts erupted on the Triassic Blomidon and Wolfville Formations which consist of red-pale green-gray, fluvial, and lacustrine, siltstone and shale (Kontak, 2005). Considerable amounts of mordenite, a zeolite, are found within this formation and the mineral name is derived from the nearby town of Morden.

The North Mountain basalts are in turn overlain by the Jurassic lacustrine limestone of the Scots Bay Formation and the red, fluvial, and lacustrine siltstone and shale of the McCoy Brock formation.

In addition to agates, 16 different zeolite minerals have been found. Chabazite, heulandite, stilbite, analcime, and thomsonite are the major species found. Originally, the zeolite assemblage here resembled the zeolite assemblage found in Iceland, formed by burial metamorphism, but has since been determined to be derived from

Fig. 7.36 Stratigraphic column Bay of Fundy basin (Hartman & Lloyd, 2012)

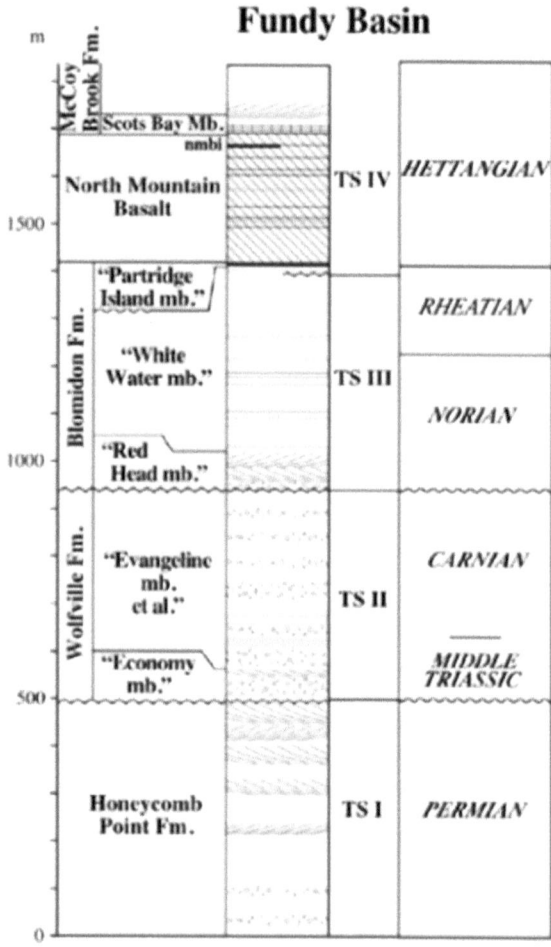

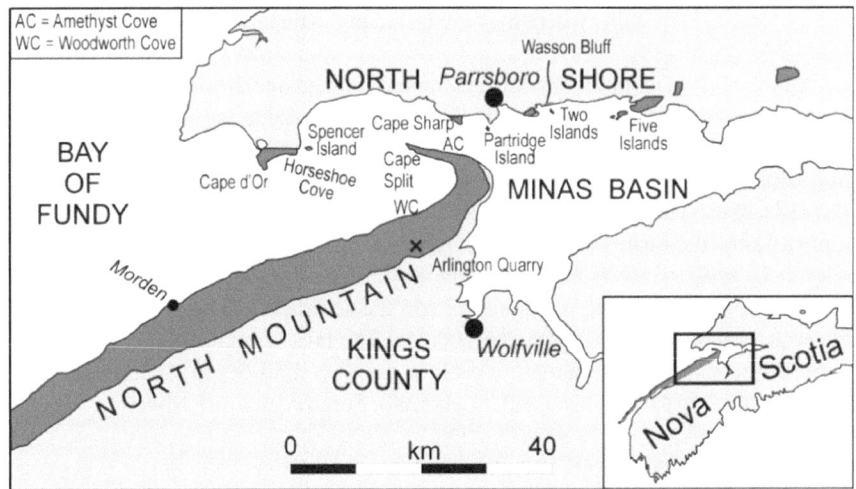

Fig. 7.37 Location of North Mountain basalt (shaded area) (Pe-Piper & Miller, 2003)

low temperature hydrothermal fluids, with chabazite forming from basaltic glass at 50–150 °C at the lower temperature scale, and heulandite forming from rhyolitic glass at 150–225 °C at the higher temperature scale (Pe-Pipes & Miller, 2003).

After zeolites, agate is the next most abundant mineral found, occupying the vesicular cavities of the middle flows of the North Mountain Basalt. The agates are transparent to cloudy white, red-brown to green in color and are oval to lobate in form. On the south shore of the Bay of Fundy, near Morden, agate is found in elongate vesicular cylindrical gas pipes or segregation pipes as pipe amygdules within the top flow of the Brier Island Member of the North Mountain basalt (Kontak et al., 2010). These agates are similar to the Potato Hill andesite agate found in West Texas. Agates are found in seven localities, primarily in the north shore of the Minas basin: Horse Shoe Cove-Cape D'Or (that has blue agate), Partridge Island, West Bay, Wasson's Bluff, McKay Head, and Two Islands. Agate is found in the south shore at Cape Blomidon along with the gas vesicle agate.

Native Copper is also found in the North Mountain basalts. This is similar to the occurrence of native copper in the Keweenaw Peninsula where low temperature hydrothermal fluids are involved.

7.9 Botswana Africa Agates

Botswana agates are found in a continental flood basalt province in the Bobonong area of Eastern Botswana; along the Motloutse River and in the Serowe area southwest of Bobonong. Every year flood waters of the Motloutse River erode the agate from highly weathered basalts, where they are found in the fields adjacent to the river (Jain et al., 2013).

With the break-up of the Gondwana Supercontinent and the opening of the Indian Ocean, the Limpopo Triple Junction was formed, located to the east of Botswana on the east coast of Africa (Jourdan et al., 2006; Manninen et al., 2008) and has a strong influence on eastern Botswana geology. It occurred in the Early to Middle Jurassic, 185–172 Mya. The break-up peaked 182 ± 3 Mya. The eastern Botswana stratigraphy, where the agates are found, is represented by the "Stromsburg Lavas" that overlie the aeolian Bobonong Formation. K–Ar dates the volcanic pile at a minimum of 181 Ma. Early Jurassic palynomorphs in the sediments suggest a maximum age of the sediments between ~195 and 200 Ma. The lavas commenced erupting about 195 Mya in the Karoo magmatic events in south-central South Africa and moved north to Botswana and Nuanetsi and Lebombo, South Africa 178 Mya (Carney et al., 1994).

The Botswana agates are found in the topmost section of the basalt flows of the Jurassic Karoo Volcanic Series. They are termed the "Stromberg" lavas in Botswana and are the equivalent of the Parana and Etendeka traps in Brazil and Namibia and the Drakensberg basalt formation in South Africa (Smith, 1984). They are widely distributed in Botswana, covering an area of 150,00 km^2, but are largely obscured by the overlying Cenozoic Kalahari beds and crop out in geographically separate areas as erosional remnants where they are assigned local formation names (Fig. 7.38). The Ramoselwana Formation, the formation of interest here, is the local formational name for the Stromberg Lava Group. Their thickness exceeds 400 m. Thirteen flows were identified in one borehole that drilled through the Ramoselwana Formation. The paucity of tuffaceous or agglomeratic beds suggests that the lavas extruded quickly and quietly. That they were subaerially extruded is indicated by weathered interflow zones as well as by the associated underlying continental sedimentary rocks and basalts of the Bobonong Formation. The Bobonong Formation consists of fluvial and lacustrine sandstones and limestones immediately overlain by thin pillow lavas, where the agates occur. The lavas erupted onto an irregular topography of eroded and partly lithified Ntame Sandstone (Carney et al., 1994).

Most exposures of the Stromsburg Lava Group consist of deeply weathered basalts that are highly amygdaloidal. The olivine-bearing basalts of the Ramelsselwana Formation consist of plagioclase, clinopyroxene, olivine, biotite, zoned Na–K feldspars and rare analcite, suggesting an alkaline and mildly undersaturated fluid affinity. The amygdules consist of the zeolites, stilbite, mesolite, and natrolite, indicating a zeolitization zone. Chalcedony, jasper, carnelian, calcite, chlorite, and muscovite are also found as amygdules in the basalt. The lavas in the lower part of the Bobonong Formation sequence contain olivine as groundmass and phenocrysts with orthopyroxene, clinopyroxene, and plagioclase. The upper part sequence rocks are olivine-free tholeiites where accessory sanidine replaces the usual glass interstices (Carney et al., 1994).

The "Botswana pink" agate occurs as amygdules and joint fillings in the lavas of the Upper Bobonong Formation. They are a distinctive gray and pink, finely banded agate and occur as nodules (Fig. 7.39). The agate is almost always coated with

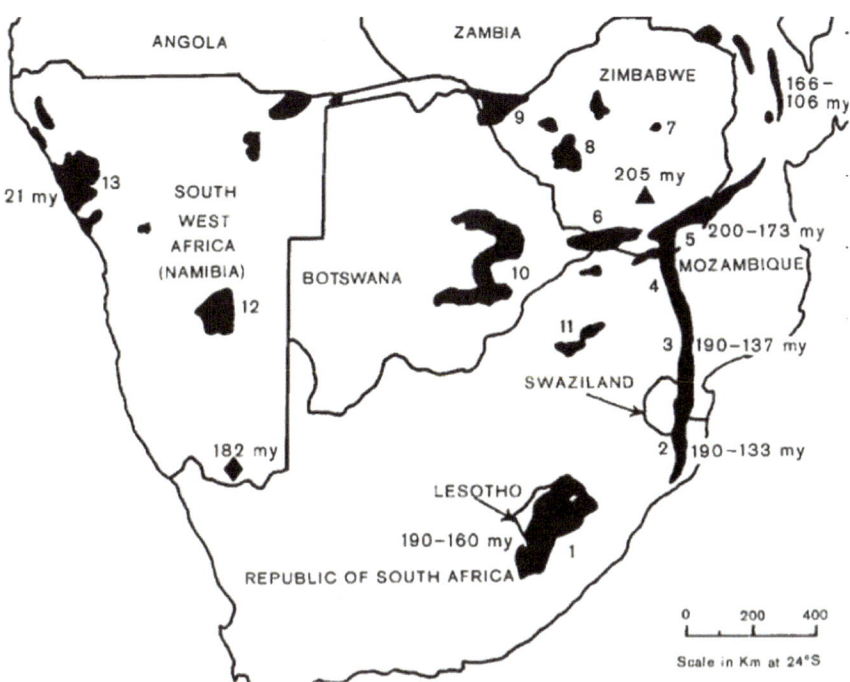

Fig. 7.38 Distribution of Karoo Lava Outcrops in Southern Africa (Bristow & Saggerson, 1983)

celadonite (Jain et al., 2013) and indicates that the area was initially propylitized, initially forming the celadonite in the vesicles of the basalt, followed by zeolitization and agatization. As the Botswana agate is an economic resource, little has been written about them and some of what was written in this writing is conjectural.

Fig. 7.39 Botswana agate;
note green celadonite areas

7.10   Ethiopia–Yemen–Saudi Arabia

The western part of the Arabian Peninsula consists of a Late Proterozoic cratonic shield area consisting of volcanic, metamorphic, and post-cratonic sedimentary rocks (Brown et al., 1989 and Fig. 7.40). Harrats (lava fields) of flood basalt erupted during the late Oligocene and early Miocene. At the same time, a continental rift valley formed along the future Red Sea axis. Tuffaceous freshwater lake beds were deposited mafic and silicic volcanoes. In late to early Miocene times, dikes, tholeiitic dikes, gabbros, and plutonic rocks intruded into the rift volcanics and sedimentary rocks. Rift spreading halted 14–14 Mya and the red sea escarpment was formed. Erosion of the escarpment deposited a coarse conglomerate. Deposition of evaporites occurred in the late Miocene. Sea floor spreading started again 4–5 Mya.

The southern shield area is a basaltic plateau formed by fissure eruptions forming the Yemen Volcanic Group (Fig. 7.40). Eruptions occurred in two phases. The first phase, and earliest phase, consists of the late Oligocene to early Miocene Yemen Trap series that are considered to be synrifting with the Red Sea Rift. The first phase occurred 31.6–15 Mya and consisted of alkaline to transitional basalts, with a few rhyolites and associated ignimbrites and tuffs. The basalts were generally a picritic basalt transitioning to an alkali-olivine basalt and Hawaiite. The second phase, the Yemen Volcanic Series, erupted 10 Mya and up to recent times and is considered to be a post-rifting eruptive phase. The outcrops are along the southern Gulf of Aden coast, in the vicinity of Sana'a, Dhamar, Marib, and the Southern Red Sea. The rocks consist of stratovolcanoes, cones, domes, and basaltic lava flows (Mattash, 2006).

Agates found in Yemen are a yellow or light orange color. The Yemen agates are found in the western part of the plateau in the basalt flows. Mattash says that there are five deposits of agates: in the Mt Elhan area, west of Dhamar and in Kholan, southeast of Sana'a. Other locations include the areas of Al-Huban and Al-Rahida, in the city of Taiz.

In Ethiopia, the Ethiopian Plateau is bounded on the east by the Afar Depression, or Afar Triple Junction (Fig. 7.40). The Depression consists of Pliocene volcanic rocks, with the axial line of the depression occupied by Pleistocene to Recent volcanics. Further east is the Danakil Horst and the Southern Red Sea rift. To the south, and forming the southern limb of the Afar Triple Junction, is the main Ethiopian Rift. The rocks are subdivided into the Trap Series and the Aden Series. The Trap Series are Tertiary flood basalts that form the northwest and southeast portions of the Ethiopian Plateau and are considered to be the counterpart to the Yemen trap series. The rocks range in age from the Paleocene to the Oligocene (30–19.4 Ma), and are up to 3 km in thickness. They overlie the Mesozoic sediments unconformably. The rocks are alkaline and are typically oxidized. In the Plateau interior are shield volcanoes that formed in Miocene times and consist primarily of alkali-olivine basalts.

In post-rifting times (Middle Miocene to Quaternary times), in Southern and Central Ethiopia, violent silicic eruptions of the Aden series covered much of the plateau, the Ethiopian Rift, and Afar Depression (the proto-rift areas). The rocks consisted of welded pantelleric tuffs (ignimbrites) and welded tuffs that covered an

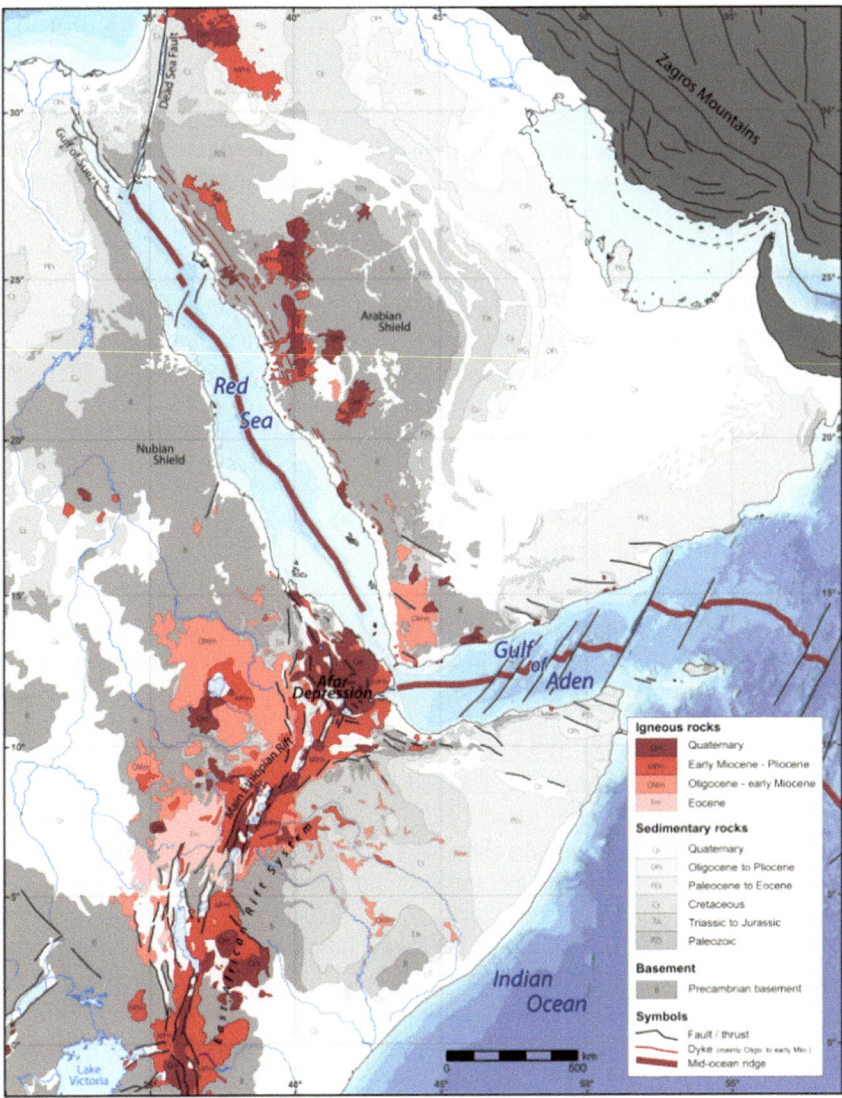

Fig. 7.40 Geologic map of the Afro-Arabian rift system and the Afar Triple Junction and Depression and showing extent and chronology of Cenozoic volcanism (after Rime et al., 2023)

area of 150,000 km^2. In the Pliocene, welded tuffs erupted along with another series of lavas, plugs, and domes. The lavas tend to be peri-alkaline, feldspathoid lavas.

Agate is present in the Ethiopian Cenozoic trap basalts, but locations of the agate are largely absent. A sole agate location has been identified at West Belessa, North Gondar in northwest Ethiopia. The marketplace advertises a blue to purple agate from Ethiopia.

Opal is found in the Wegel Ten Plateau within the Ethiopian trap rocks (Chauvire et al., 2019). The opal is found in a weathered rhyolitic ignimbrite layer that is a few meters thick and 100s of meters wide. The opal is embedded between unweathered layers of ignimbrite. No traces of fluid circulation or subsequent weathering are observed in the unaltered thick welded layers above the opal layers. This constrains opal formation to the period after its host rock deposition and before the overlying volcanic strata. The opal layer and the overlying layer were age dated at 30.6 Ma and 29.4 Ma, respectively, or the Oligocene period, and indicate a 1.2 Ma period for the formation of opal. The opal was formed by continental weathering processes, that is, supergene weathering and dissolving of the ignimbrites. Both the glass and feldspars of the ignimbrite are the source of the silica. With the release of silica, opal A is formed. The oxygen isotope signatures, $\delta^{18}O$, ranged from 26.52 to 30.98 0/00 versus SMOW and indicated temperatures of formation to be 18–21 °C (Chauvire et al., 2019).

7.11   Khur Area, Iran

The Khur agate province is located in northeast Ishahan province in the central Iranian plateau and is located on the Central Iranian plateau/microplate (Fig. 7.41). With the collision and subduction of the Afro-Arabian plate beneath the Iranian Central Continental margin plate during Eocene-Miocene times, the Central Iranian magmatic arc was formed. This Eocene volcanism started as continental arc magmatism with accompanying subsiding basins that continued into the Oligocene–Miocene times. The magmatic arc (the Sakand-Bazman Volcanic belt) consists of subalkaline volcanic basalt, andesite, and rhyolite that was formed in the Central Iranian plateau and peaked in Eocene times. Later thrust faulting and block faulting subdivided the microplate into horst and graben blocks.

The Khur agate area is in the Tash Tab Mountains and in the Yazd block in a graben surrounded by three major thrust faults (Fig. 7.42). It is on the on the northwest slope of Howz-e-Mirza Mountain, where Eocene volcanic rocks outcrop (Fig. 7.43). The rocks consist of alkali basalt, trachybasalt, trachyandesite, and trachyte. The Qom formation, consisting of sandstone and limestone, is the top of the stratigraphic interval, with the Darreh Aizir conglomerate at the base. The volcanic units form small hills in the area and are surrounded by extensive bentonite beds. Agates and geodes are found in a basal tuffaceous vesicular andesite (Nazari, 2003). The tuffaceous andesites grade up into thin interbeds of tuff and vesicular andesite and dacite lavas. The basal andesite flow is 30–200 m thick, but 100 m thick at the agate locality. The agate stratigraphic interval has numerous fissures, with pyroclasts in a tuffaceous ashy matrix, interbedded with vesicular andesite beds and extensive bentonite beds that are currently mined. Numerous faults, trending northeast to southwest and east–west are in the area, with the agates themselves showing evidence of the faulting. In addition to agate and geodes, jasper, calcite, hematite, dolomite, and quartz are found. Siliceous veins occur near the faults and joints.

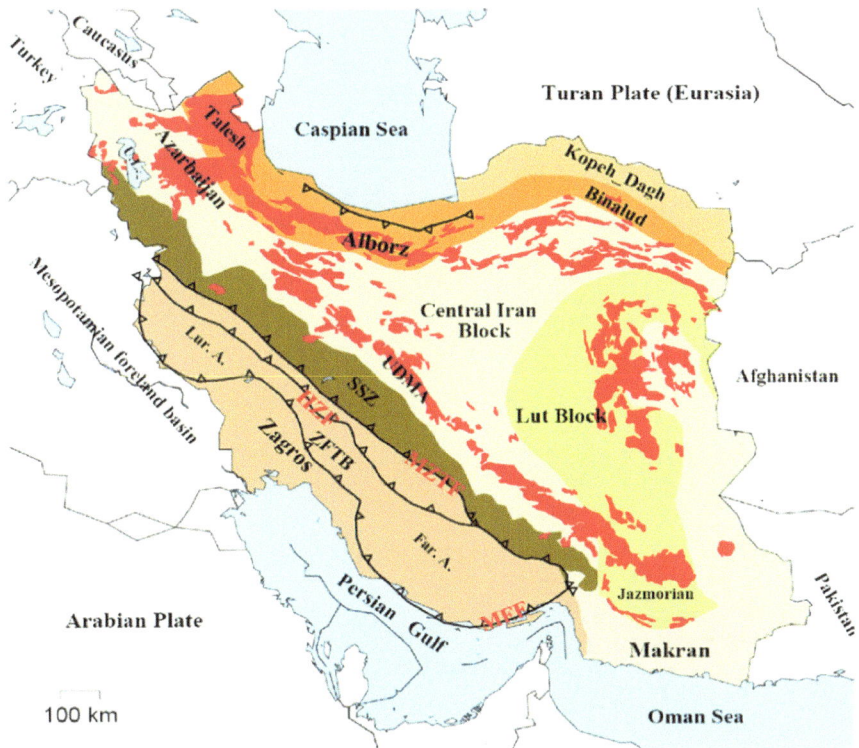

Fig. 7.41 Central Iranian microplate and Cenozoic vulcanism (After Tazhizadeh-Farahmand et al., 2014)

The bentonite consists of montmorillonite, kaolinite, quartz, calcite and cristobalite and is an alteration product derived from volcanics, usually volcanic ash. Mahmoudi et al. (2013), through the analysis of the bentonite and the application of a discrimination classification diagram, plotting Zr/TiO_2 versus Nb/Y values (as proposed by Winchester & Floyd, 1977), indicated that the bentonites were derived from an andesite-basalt parent. Oxygen and Deuterium isotope values ($D^{18}O$ vs. SMOW and D^2H vs. SMOW) indicated temperatures of formation at 83 °C for the formation of the bentonite and were formed in comingled hydrothermal and meteoric waters.

Fluid inclusions in the agates were single phase and indicated temperature of formation at less than 100 °C. (Mahmoudi et al., 2013). Calcite in the agate suggests that siliceous hydrothermal fluids intermingled with alkaline basinal waters, forming at temperatures under 200 °C. (Mahmoudi et al., 2015). The rare earth elements (REE) composition of the andesite host rock contained high concentrations of Ce, U, and B and was also found in the agate at lower concentrations, as well, suggesting a silica source from hydrothermal fluids (Mahmoudi et al., 2015). A lack of water in the

Fig. 7.42 Location map of the Khur Graben agate area (Mahmoodi et al., 2013)

agates prevented any $\delta^{18}O$ temperature estimations, but temperature is estimated to be from hydrothermal waters with an atmospheric source (Mahmoudi et al., 2015).

The Khur agates, according to Mahmoudi, formed in the cavities of the Eocene andesitic basalt within the bentonite beds and were formed before the faulting and jointing movements. XRD also identified dolomite, calcite, and barite in the agates. Agates, then, formed from the periodic eruptions of low temperature hydrothermal fluids. The oxygen and deuterium isotope data indicate hydrothermal fluids commingled with fluids of an atmospheric origin.

Bentonite is usually an alteration product from volcanic glass (Hauser & Reynolds, 1917; Slaughter, 1963, etc.). The silica source for the agates in the Khur area is most likely the volcanic ash that weathered or altered to bentonite and released silica (Fig. 7.44). In addition, the volume of silica needed for the number of agates found here would be provided by the volume of bentonite found here. The agates then formed from low temperature fluids from late phase magmatic waters commingled with meteoric waters that altered volcanic glass and released silica.

The agate types include faulted or fractured agates, manganese oxide plume agates, sagenitic agates (pseudomorphs after aragonite), flame, and pompom agates. The agates are similar to the West Texas plume agate. The outer surface or rinds

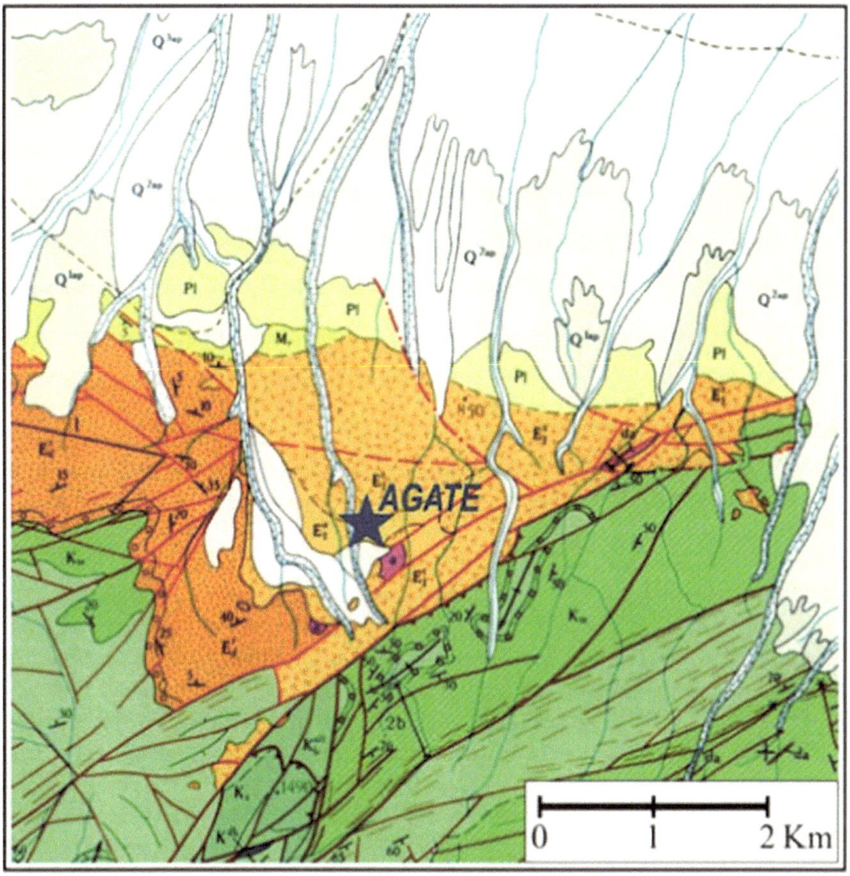

Fig. 7.43 Geologic map of the Khur Area; orange-brown are Lower Eocene volcanics (andesite, dacite, and tuffs); green are Cretaceous sediments (Nazari, 2004)

Fig. 7.44 Khur agate area, bentonite over agate-bearing andesite (Nazari, 2004)

of the agates are generally composed of calcite, siderite, limonite, and clay. Agates are pale to inky blue, but some are gray, white, colorless, and rarely red. Cesium causes the violet color found in some agate. Inclusions of aragonite, goethite, calcite, siderite, and pyrolusite are common. Other associated minerals found include celestite, hematite, celadonite, and anhydrite. Geodes consist of colorless euhedral quartz with occasional amethyst.

7.12   India

Indian Deccan traps flood basalt province (Fig. 7.45) occurs at the Cretaceous Tertiary Boundary (Alexander & Paul, 1972). Currently, the northward-moving Indian plate is colliding with the Eurasian plate, creating the Himalayan Mountains. During Cretaceous/Tertiary times, seafloor spreading occurred at the boundary of India and African plates that pushed India over a mantle plume/hotspot named Reunion.

The Deccan flood basalts erupted from fissures 67.5–60.5 Mya in three main phases. Phase I erupted ~67 Mya. Phase II erupted ~65 Mya. 80% of the Deccan traps formed at this time, all within a 500,000-year timeframe (Krishnamurthy, 2020). Phase III erupted ~64–60 Mya mainly in the southern part of the Deccan igneous province. The present observable extent of the Deccan flood basalt is 512,000 km^2. The basalts are mainly quartz and hypersthene tholeiites in the plateau regions of Western Ghats, Malwa and Mandla. In the rift zones of Narmada-Tapi, Cambray, and the Saurashtra and Kutch areas, however, there are dyke swarms, intrusive and extrusive centers, and ring complexes composed of alkaline rocks, rhyolitic rocks, picritic basalts, basanites, among others. The flows are basically two types, a compound flow (pahoehoe) and a simple flow (aa). The compound flow is vesicular at the top and has pipe amygdules at the base and a dense core. Ropy structures, squeeze-ups, and spheroidal amygdules are common. Intercalated beds of clayey and marly sediments are common. The surface of the traps is weathered and denuded during Tertiary times and a ferruginous laterite or bole was formed. The basalt is occasionally jointed in places and weathers to rounded to subangular blocks.

In addition to agates, which are seemingly ubiquitous throughout the Deccan igneous province, the traps are primarily known for their secondary mineralization, particularly the zeolites; especially heulandite, stilbite, merlinoite, and analcite. Various classification and zoning schemes have been proposed (Avasia & Gangopadhya, 1984; Oettens, 2003; Sukheswala & Poldervaart, 1974; Sukheswala et al., 1974 etc.) and not substantiated. The most recent, and probably the most accurate, Ottens et al. (2019), has shown a three-stage classification of the secondary mineralization that occurred in the Deccan igneous province (Fig. 7.46).

In the earliest Stage 1, the phyllosilicates and clay minerals formed, e.g., celadonite and smectite. In Stage 2, calcite, generations I and II, zeolites generations I and II, and chalcedony formed, e.g., stilbite II, plagioclase IIb, heulandite IIb, mordenite IIb, and silica IIc. Occasionally the mordenite formed in silica. Stage 3 consists of calcite generation III, the zeolites, powellite, apophyllite (from late-stage hydrothermal

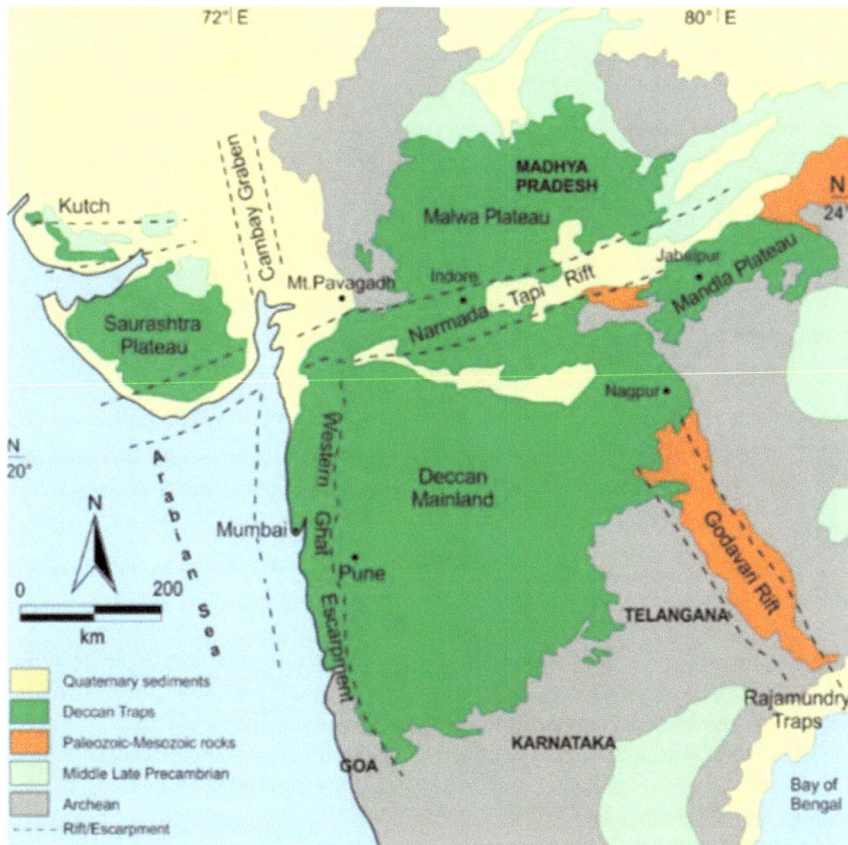

72° E   80° E

Fig. 7.45 Geologic map of the Indian Deccan Flood Basalt Province (Krishnamurthy, 2020)

fluids), e.g., pavellite IIIb, c, and apophyllite IIb, c. Note the presence of celadonite, a green chloritic clay mineral, found associated with agate rims.

The chemical elements necessary for the secondary minerals were derived from the alteration of the host rock, e.g., volcanic glass, olivine, clinopyroxene, and plagioclase. $\delta^{18}O$ and $\delta^{13}C$ data indicated that the early phases of the mineralization involved primarily magmatic waters; with the later phases involving increasingly greater amounts of meteoric water. Rb–Sr and K–Ar dates obtained from apophyllite indicated that crystallization occurred over a long period of time, from the Paleocene to the early Miocene, with clusters at 55–58, 44–48, 25–28, and 21–23 Ma.

Agates from the flood basalts of India have not been studied much, with the studies concentrated mostly on zeolites. What studies there are (Randive et al., 2019) present a very localized picture of agate occurrences in the Deccan tap rocks. This is a similar picture of agate occurrence as in the Ethiopian flood basalt province. In addition, the Indian agates present an uncanny appearance to the Montana moss agate, suggesting similar modes of origin. Agates, yellow and black-skinned, and resemble Montana

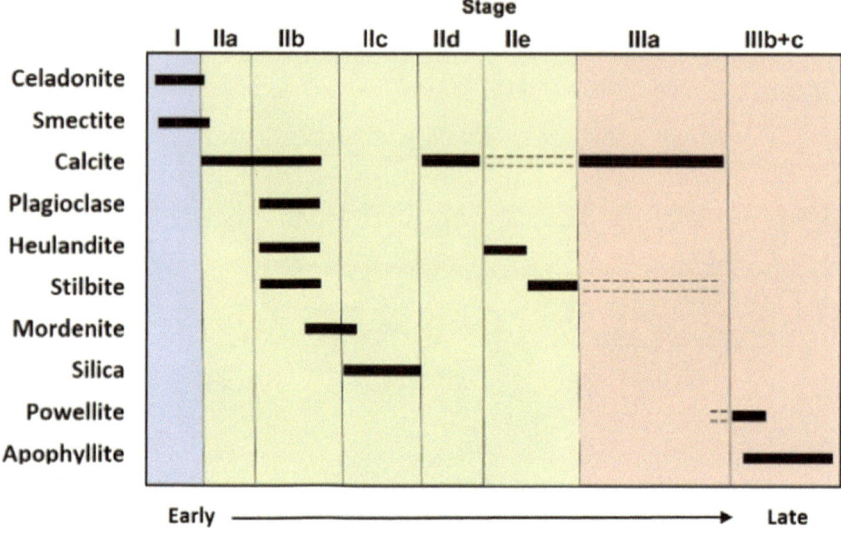

Fig. 7.46 Secondary mineral classification in the Deccan traps (Ottens et al., 2019)

moss agates (Fig. 7.47). In addition to the vesicular, amygdaloidal agate, agates are also found in the Tertiary conglomerates, of the Gaj and Siwalik Formations/Group. The agate mines of Ratanmal in the Rajpipla, Gujarat State mine agates from this conglomerate. The agates are also mined from the surface ferruginous laterite. Agates come from other various states and districts on or near the edge of the traps, but most come from Rajpipla.

Spherulites and thundereggs are also found in the igneous province as described by Kshirsager et al. (2012). They were found in the northwestern area of the igneous province, of Saurashtra. Some silicic rocks, necessary for the formation of spherulites

Fig. 7.47 Indian Black Skin agate

and thundereggs, are found in this area, namely, rhyolites, granophyres, trachytes, hyaloclastites, tuffs, pitchstone or ignimbrites, and volcano-sedimentary rocks. Data indicated that these spherulites and thundereggs formed primarily from meteoric waters commingled with some late-stage magmatic waters.

7.13 Germany Variscan Orogeny Agates

At the end of the Carboniferous period, the assembly of the Pangea supercontinent was almost completed with the last phases of the Variscan or Hercynian orogeny and the final closure of the Iapetus Ocean, forming the Rheno-Hercynian or Variscan Mountains (Fig. 7.48). An east–west foreland basin was created in front of the Variscan Mountains from the United Kingdom to Poland. Crustal extension created a Basin and Range type geological province similar to the Basin and Range geological province in the western United States of America and Northern Mexico. The Saar-Nahe basin in southwest Germany is a post-collisional Variscan orogenic intermontane basin, where thick volcano-sedimentary deposits were rapidly buried (Fig. 7.49). At the northern margins of the orogenic belt, a southward dipping, suture detachment fault created the present-day Saar-Nahe basin as a half graben.

At the end of the Variscan orogeny, 292 Mya in the Lower Permian, volcanism produced maars-type volcanoes, diatremes, sills, laccoliths, and lava flows in an intermontane basin. This volcanism began late in the basin, beginning after 5,000 m of continental sediments were deposited. The volcanism created a series of lava flows,

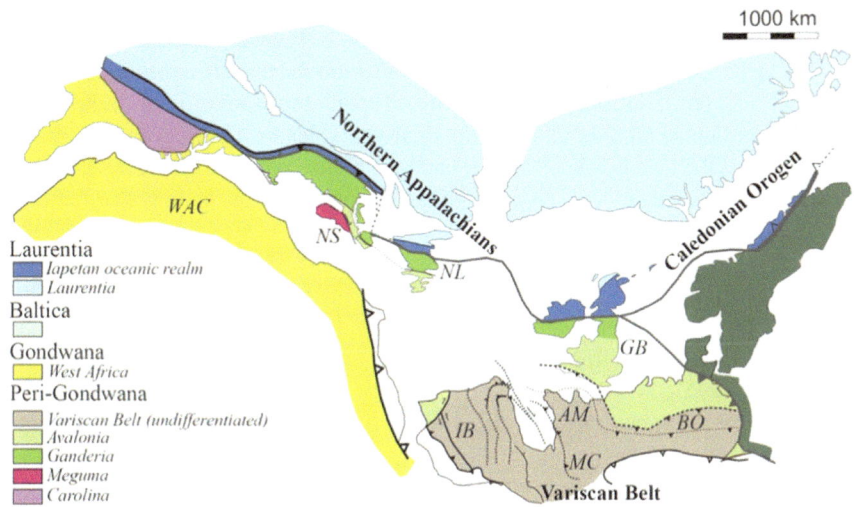

Fig. 7.48 Location of the Variscan orogenic belt at the end of the Paleozoic (After Henderson et al., 2018)

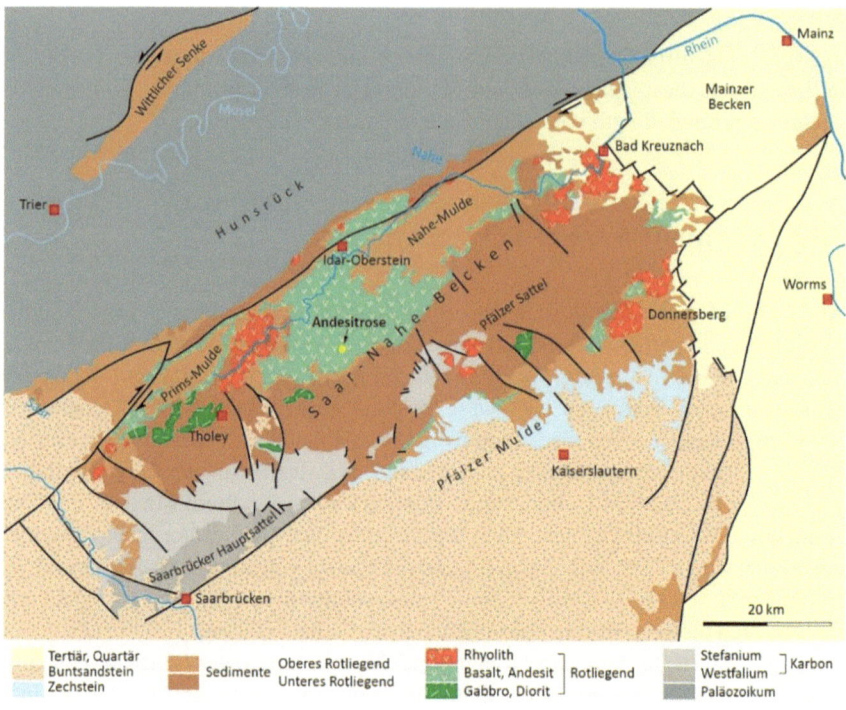

Fig. 7.49 Geological sketch map of the Saar-Nahe basin (DGadmin, 2022)

800–1,000 m in thickness. These are the effusive volcanics of the Roetliegand group, Nahe subgroup, Donnersberg formation (Lorenze & Haneke, 2004).

The lavas are restricted to the Donnersberg formation and are considered synsedimentary with sediment deposits, that is, they erupted at the same time as the sediments were being deposited. About 26 maars diatreme volcanoes have been identified as the source of the lava flows. However, late Paleozoic erosion has removed the tephra rings and maars volcanoes, leaving only the diatreme volcanoes. In general, though, magma vent and feeder dyke locations are unknown. There are no scoria cones. Also, a possible shield volcano has been identified north of the Saar-Nahe basin. Three lava flow series separated by 6 tuff/ash marker beds have been identified. These basal lava flows of Lava 1 are a tholeiitic basalt to a basalt andesite and are 15–40 m thick. As the flows grade upward to Lava 2 to Lava 3, their composition becomes more intermediate to acidic, andesitic to rhyolitic, and are up to 200 m thick. As per composition-wise, the tholeiitic basalts and basaltic andesites could qualify as a "flood basalts."

The lower part of flows has some recognizable pipe amygdules. The upper portion of the flows is vesicular. Early Permian weathering had destroyed much of the original surface textures. Only one pahoehoe surface has been identified. Lava clasts in one overlying sediment layer indicated that the lava flow was basically a subaerially emplaced pahoehoe lava flow. However, most of the lava flows are considered to be an aa flow.

Above the Donnerberg formation, in the Glan subgroup, are distally located, ash and tuff beds intercalated with lake sediments. These ash and tuff beds erupted from a source outside of the Saar-Nahe basin, most likely the Velding rhyolite plug, located north of the basin, and were subaqueously deposited and reworked. The sediments represent deposition in a shallow, lacustrine and fluviatile environment. Pumice lapilli are compacted and welded and represent post-sedimentary diagenetic compaction. Ignimbrites emplaced during Glan times have been identified in the Prims syncline and the Nahe basin. Their source is probably from the Velding rhyolite plug also.

Groundwater and meteoric water/rainwater enter the joints and fissures of the flows and cause thermohydraulic explosions. This elevated heat flow of water has its origins in the emplacement of an upper mantle magmatic reservoir located beneath a thinning crust and the subsequent expansion of these magmas into the basin. The presence of groundwater in the sediments easily conducts the heat and results in diagenesis and alteration of the rocks at a relatively high temperature with the formation of secondary minerals: albite, chlorite, calcite, epidote; the zeolites, prehnite and pumpellyite; hematite, and kaolinite ((Lorenze & Haneke, 2004).

Agates are found in the vesicular andesitic rocks at Idar-Oberstein. Idar-Oberstein is a prolific agate mining area that has historical records dating back to 1497 (Fig. 7.50). In addition to the reddish-pink to light pink colored agate (some with celadonite skins), there is smoky quartz, amethyst, jasper, and calcite. Most mines or quarries are closed. Mine/quarry at Galgenberg/Steinkaulenberg is the only mine open. Closed mines/quarries include Jurhem quarry, Bernhardt quarry, Vollersbach quarry, and Setz quarry area of Fischbachlal Valley, Saarland.

Further east along the same lower Permian Variscan orogenic belt, in the Thuringian Forest Basin, thundereggs are found. These thundereggs are described by Holzhey (2013) (and in the thunderegg section of this writing), formed in the Oberhof quartz porphyry rhyolite of the Lower Rotliegend group.

Further east, yet, in the NW Saxony Basin, are also found also thundereggs (Fig. 7.51). Unlike Holzhey, Götze (2023) describes these thundereggs as agates. Like other thunderegg deposits, they are found exclusively in rhyolites and ignimbrites. In the NW Saxony basin, they are found in the Lower Permian glassy component of the Kemmlitz rhyolite. The Kemmlitz rhyolite is underlain by the Rochlitz ignimbrites and tuffs and is overlain by the Wurzen formation that consists of a caldera system consisting of stratovolcanoes, ignimbrite flows, and intrusives (Fig. 7.51). The thundereggs are found in spherulites or lithophysae. The rhyolite and spherulites have been completely altered from sanidine, orthoclase, and quartz to kaolinite, illite, and smectite clays during the late Cretaceous to Miocene times. The mineralogy of the rhyolites and spherulites indicates a high temperature of formation accompanied by rapid cooling. The Kemmlitz rocks are dated to 289.8 Ma and the Wurzen is dated to 289.3–287.3 Ma; indicating an agatization time of 0.5–2.5 Ma process.

Götze's fluid inclusion studies of the Saxony agate indicate temperatures of formation from 134 to 186 °C with the mobilization and accumulation of silica into the spherulites or lithophysae during late volcanic or hydrothermal activity or soon after.

Fig. 7.50 Postcard of Oberstein (1907), by Ludwig Feist, showing the Nahe River, town of Oberstein and agate-bearing andesite outcrops

Further, rare earth element (REE) and heavy rare earth element (HREE) indicate interactions of the host rocks (rhyolites and ignimbrites) with SiO_2, magmatic volatiles (F, Cl, and CO_2), and meteoric and surface waters. Cathodoluminescence (CL) showed a characteristic yellow CL. Internal textures and a high defect density of macro- and microquartz detected by electron paramagnetic resonance (EPR) spectroscopy indicated a crystallization from an amorphous silica precursor under non-equilibrium conditions.

7.14 Poland Variscan Orogenic Belt Agates

Early Permian Roetlingland sedimentation in the intermontane basins of the North and Intra Sudectic basins was sometimes accompanied by intense volcanic activity of two distinct magmatic suites: (1) rhyodacites, andesites, and basaltic andesites were characteristic of the earlier stages and (2) basaltic trachyandesites, trachyandesites, and rhyolites erupted during the later stages.

Agates mainly occur in the Lower Permian Roetlingland basalts and andesites of the North-Sudetic and the Intra-Sudetic Depressions (Powolny et al., 2019 and Fig. 7.52). They are in the vicinity of Płóczki Górne, Wleń, Lubiechowa, Różana, Czadrów quarries near Kamienna Góra, Suszyna, and Niwa. Thundereggs are found in rhyolites and tuffs near Nowy Kościół, Gozdno, and Sokołowsko. Less known are agates found in basalts in the vicinity of Alwernia, Regulice, and Rudno near

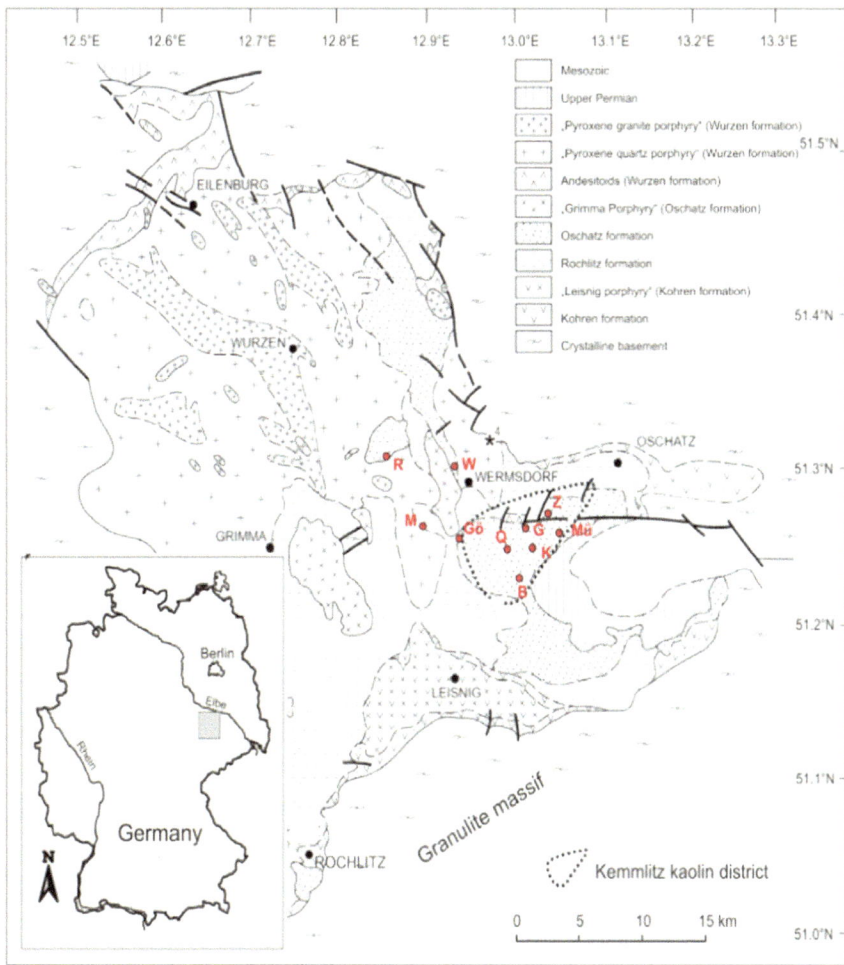

Fig. 7.51 Geologic map of NW Saxony basin showing agate/thunderegg locations by Götze et al. (2023)

Krakow. They are thought to originate from post-magmatic hydrothermal fluids of low temperature. Inclusions in the agate include celadonite, plagioclase, hematite, goethite, barite, calcite, and the zeolites, heulandite, clinoptilolite, nontronite, and saponite. Textures suggest the hydrothermal low temperature regime for the formation of the agates: colloform, comb, feathery, and puzzle jigsaw (Powolny et al., 2019).

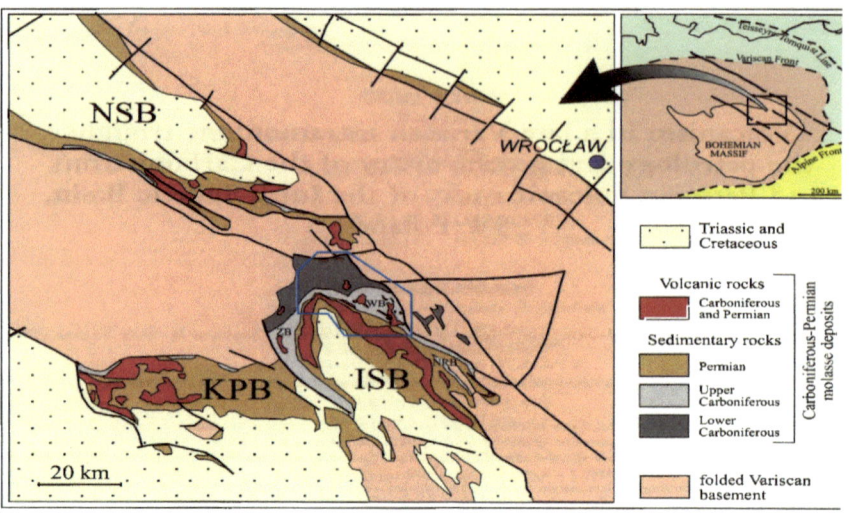

Fig. 7.52 Geologic map of the North and Intra-Sudetic basins, Poland (Dumanska-Slowik et al., 2018)

7.15   Scotland

7.15.1   Midland Valley

The Northern Midland Valley of Scotland is an ancient rift zone consisting of grabens with basalt and andesite lava flows belonging to the Old Red Sandstone group that is Devonian in age; roughly 410 Ma in age (Browne et al., 2002). The Midland Valley rocks are associated with the Caledonian/Acadian orogenic episode that produced the Laurentia supercontinent in Devonian times (Figs. 7.48 and 7.53). Graben development in a sinistral strike-slip movement or regime started in the Late Silurian. Clastic infill to the grabens was punctuated by calc-alkaline eruptions from composite or strato-type volcanoes. Alluvial fans and fluvial sandstones of continental origin are intercalated or interstratified with the calc-alkaline lavas and volcanoclastic rocks. The subaerial volcanic eruptions proceeded from at least four centers and are locally eroded. These rocks are collectively called the Old Red sandstone: divided into the Lower Old Red sandstone of Lower Devonian age, separated by an unconformity in the Middle Devonian, and the Upper Old Red sandstone of Late Devonian age. The intercalated lavas of the Lower Devonian make up the Ochil Hills Volcanics member in the Sterling Perth Dundee region and are dated to 407 ± 6 and 410 Ma (Fig. 7.54). In the Stonehaven-Fettercairn region, we have the Tremuda Bay, Crawton, and the Montrose Volcanic members that initially produced thick piles of lavas accumulated locally from composite or strato-volcanoes, culminating into thinner, but still widespread, ignimbrites and pyroclastic deposits (Broome et al., 2002).

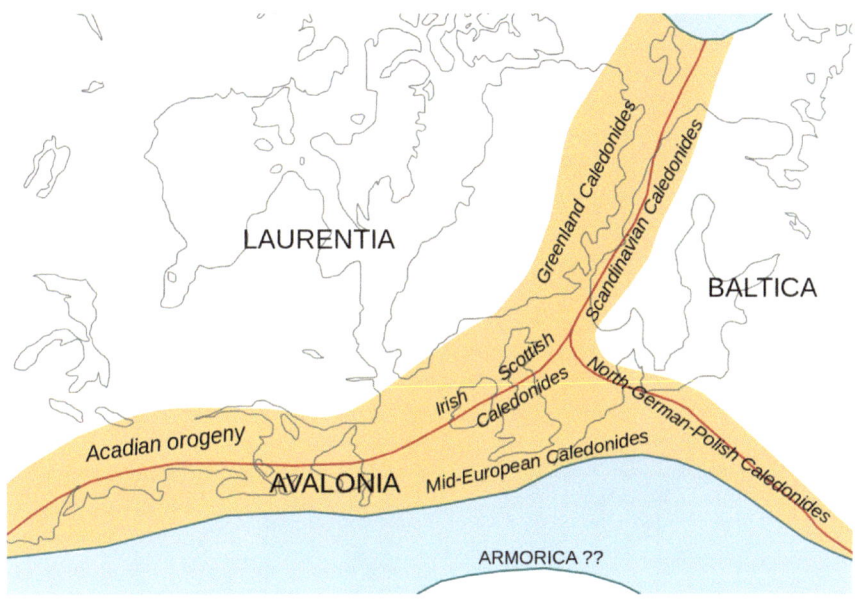

Fig. 7.53 Position of the Caledonian orogenic belt in the Devonian period (Wikipedia, 2024)

Agates are found in the basalt–andesite flows of the Ochil Volcanic formation in the Midland Valley. The most prolific and colorful agates are found in the vesicles of these basalt and andesite lava flows in the Midland Valley graben terrain along the southern margins of the Highland Boundary Fault: the Montrose and Dundee area, the Perth and Fife area, the Ayrshire area, and the Arran and Kintyre areas. The Cheviot Hills agates are found in a different area, south of the Midland Valley, in the Southern Uplands area. The Cheviot Hills area is a plateau of Lower Devonian andesitic lava flows, the Cheviot Hills Volcanic formation, punctuated by a granitic intrusion or pluton. The lavas are contemporaneous with the lavas of the Ochill Hills Volcanic formation in the Midland Valley to the north.

The most significant agate sites are the Bluehole at Usan and the Ardownie Quarry near Monifieth, which are accordingly described.

7.15.2 Bluehole at Usan Near Montrose Tayside

The most celebrated Scottish agate locality is the Blue Hole at Usan, near Montrose, Tayside. This is the agate site discovered by the "legendary" geology professor Matthew Forster Heddle (1829–1897). The stratigraphy of the area consists of rocks of the Arbuthnot-Garvock Group of lower Devonian age (Browne et al., 2002). The

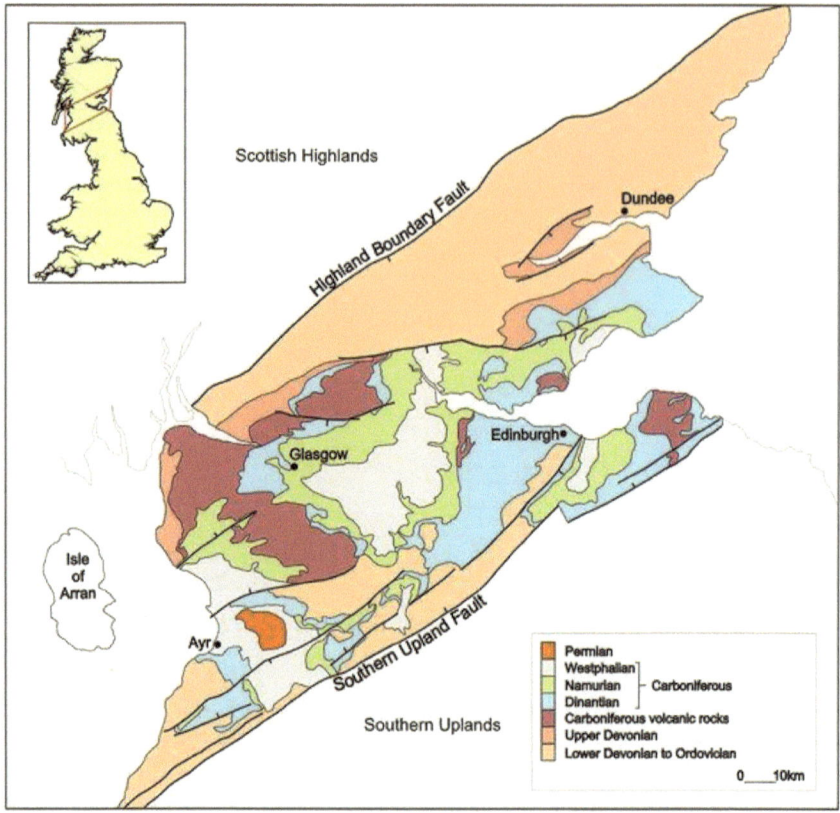

Fig. 7.54 Simplified geologic sketch map of the Midland Valley, Scotland (after Whitbread et al., 2014)

basal Dundee formation is a cross-bedded sandstone of fluvial origin with many inter-calated flaggy, thin-bedded sandstone, siltstone, and mudstones of deltaic and lacustrine origins. Overlying the Dundee formation are the lava flows and volcanic detrital rocks of the Ochil Volcanic formation. The Ochil Volcanic group consists of lava flows and intercalated volcaniclastic sedimentary rocks, mainly conglomerates. The lava flows are predominantly andesitic, though some basaltic and even rhyolitic flows are present. The formation passes laterally and upwards into the Scone Sandstone formation. The Scone sandstone consists of conglomerates and pebbly sandstones with volcanic clasts where the volcanic strata underly the unit. The sedimentary rocks are interpreted as fluvial arenites of a braided river system with the source area of the sands from a positive area located to the east, where the present coastline is situated. Lacustrine deposits formed from impeded drainage caused by volcanic piles. Nodular pedogenic carbonate is present in over bank deposits reworked by meandering streams in a hot, seasonally wet, and dry climate.

The upper surface of the andesite flows is vesicular and filled with amygdaloidal agate. In places the flows are scoriaceous. Deep fissures in the flows are filled with a green sandstone, the so-called "Neptunian" dykes. The agates are predominantly a dark inky blue, though a pink variety is also found. Secondary minerals found here include calcite, barite, goethite, saponite, celadonite gypsum, and the zeolites analcime, stilbite, and natrolite.

7.15.3 Ardownie Quarry

Ardownie Quarry agates were discovered in 1992. The Quarry, operated by Geddes Ltd, is approximately 2 km north of Monifieth, near Dundee, in the Tayside region, with access via the A92 Dundee to Arbroath Road.

Agates are found in the amygdaloidal section of andesitic lavas of the Ochil Volcanic Formation overlain by sandstones of lower Devonian age dated to c. 400 Ma. The flow is up to 140 m thick, extruded onto the floodplain of a major river system that flowed to the south-west. The sequence is part of the east limb of the Sidlaw anticline that dips to the south-east at approximately 10°. Numerous "Neptunian" dykes, which are vertical fissures in-filled with quartz-rich sediment, up to 20 m deep and 2 m wide, cut the andesite lava flow. The topmost 3 m of the andesite flow is auto-brecciated (shattered from upwards escaping gasses) and filled with quartz-rich sediment. Below the auto-brecciated zone are approximately 10 m of amygdaloidal andesite filled with agate. Due to the dip of the rocks and post-glacial erosion, exposures in the south wall of the quarry show only the upper portion of the andesite flow. Field evidence from the quarry indicate that the andesite is part of a single continuous lava flow of at least 45 m in thickness. The amygdaloidal zone contains agates from weathered andesite that are red-orange-yellow-brown due to ferric (oxidized) iron. Agates from unweathered andesite are typically blue-gray-black due to ferrous (reduced) iron. Agates are most commonly in the 10 × 5 cm size range and can reach up to 60 × 30 cm in size. Except for the south wall of the quarry, the upper auto-brecciated and the agate-bearing amygdaloidal zones have been removed by erosion and agates are subsequently found in the overburden in the vicinity of the Quarry.

Associated minerals include barite that are in quartz-lined geodes, often accompanied by calcite. Calcite occurs in many amygdules, either as total in-filling or as partial in-filling, together with quartz or agate or both, with no deposition sequence apparent as calcite may form at the center of the agate or in the periphery of the agate. Nailhead and dogtooth calcite crystals in geodes occur less abundantly than in the amygdaloidal calcite. Celadonite occurs as a dark green coating on the outside of amygdules and also as filaments in the chalcedony to form moss agates. Goethite is included in or penetrates the quartz crystals in geodes. Hematite, together with quartz and calcite is very infrequent as reddish-brown botryoidal layers in the larger geodes. Quartz occurs as euhedral crystals in the geodes and centers of the agates, frequently as amethyst and smoky quartz.

7.15.4   Skye, Mull Island, Rum, Arran

Pangea began to break up during the Jurassic period, and Britain drifted north on the Eurasian plate. Between 50 and 60 Mya, the Atlantic Ocean began to form, separating Scotland and North America. The event caused large-scale volcanic activity in Scotland.

In the early Paleocene between 63 and 52 Ma, the last volcanic rocks in the British Isles were formed. As North America and Greenland separated from Europe, the Atlantic Ocean slowly formed. This led to a chain of volcanic sites west of mainland Scotland including the Inner Hebridean Islands, Skye Arran, Mull, and Rum (Bell & Willamson, 2003; Emeleus & Bell, 2005).

The magmatism early in the Tertiary period resulted in the eruption of extensive plateau lavas (Skye, Small Isles, Mull), of a Central-Group lavas (Mull, central Arran), emplacement of central intrusion complexes (Skye, Rhum, Mull, Blackstones Bank, Arran), and of sill complexes (northern Skye, Shiant Isles, western Mull, southern Arran) (Fig. 7.55). Later volcanoes postdate the lavas, with calderas located on Skye, Mull, Arran, and Rum Islands. These volcanoes were siliceous in composition and produced ignimbrite flows, ash flow tuffs, air-fall ash and other pyroclastics (Gooday et al., 2018). Later glaciation modified the islands' surface.

The host rock for the agates are the lava flows composed of basalt. Many of the agates are found on the beaches, having been eroded from the host rock. The agates are mostly a distinctive gray-blue color, but a cream-colored agate is also found. Calcite and the zeolites, chabazite and natrolite, are also present as amygdules. The most productive locations for these agates are on the islands of Rum Iona, Mull, and from the western cliffs of Skye Island.

7.16   Australia Agate

The Queensland agate is found in the Northern Queensland Agate Creek "fossiking" area, 70 km south of Forsayth, where it is located in the western part of the Forsayth subprovince of the Proterzoic Etheridge province of the North Australian craton (Morrison et al., 2019). The region is an area where there is a long history gold mining; the gold is found in epithermal rhyolite deposits.

In the late Permian to Late Triassic in a subduction zone, intrusives and continental volcanic rocks were extruded. An extensional phase in the Late Triassic produced bimodal and alkalic magmatism. The Agate Creek volcanic center is located on the Robertson River fault, 30 km north of Gilbertson and consists of an elongate oval-shaped block of c. 275 Ma old Permian intrusive, extrusive volcanic and volcaniclastic rocks, 12 × 6 km in size, and is up to 1,000 ft thick. The volcanic sequence consists of rhyolitic ignimbrite and basaltic andesite that has abundant agate amygdules. The andesite flow (where the agates are found) is known as the Black Soil Andesite and is part of the Agate Creek Volcanic Group, which includes the Big Surprise Tuff,

Fig. 7.55 Distribution of
Palaeogene volcanic central
complexes, lava fields, sill
complexes, and dyke swarms
in Western Scotland and
northeast Ireland in the
Western Isles of Scotland
(Emeleus & Bell, 2005)

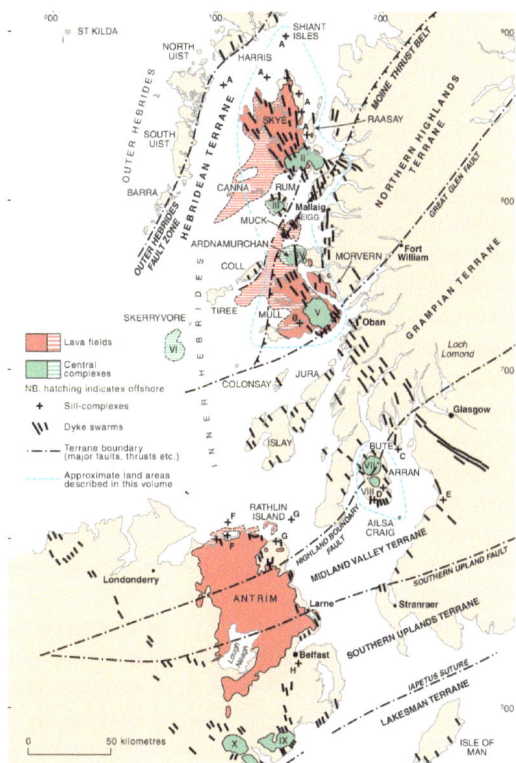

Thunder Egg Rhyolite, and the Robin Hood Granodiorite. These rocks are overlain
by the Middle to late Jurassic Connie May dolerite lava flow and capped or covered
by the Jurassic Hampstead Formation, which consist of sandstones and conglomerates, derived from the erosion of the underlying volcanic rocks. The sandstones are
highly arkosic and are thought to be the source of the agate silica for the agates. The
Hampstead sandstone is a fine to medium quartzose sandstone with some pebbly
and conglomerate beds, is strongly cross-bedded, and is interbedded with micaceous
siltstone, mudstone, and a very fine-grained sandstone. The Hampstead formation
represents a fluvial environment with fossiliferous channel and floodplain deposits
(Smart & Senior, 1980). The Permian rocks are separated from the Jurassic rocks by
an erosional unconformity.

Agate was first found in the area about 125 years ago and is found at the head of
three creeks: the Blacksoil Creek, Spring Creek, and Agate Creek (Howard, 1996).
The area is 26 km in length, with the creeks flowing northwest into the Robertson
River. The agate area is currently 11 × 3 km in area. Agate is in found as amygdules
in the vesicular upper parts of the Black Soil Andesite and is also found in the nearby
alluvial gravels; the andesite having undergone extensive weathering. Agates occur
as nodules (solid agate) or as geodes, roughly ellipsoidal or rounded in shape in

various sizes, ranging from pigeon egg size to football size, but averaging about 50–70 mm in size. The main localities are Flanagan, Simpson, Black Rock, The Saddle, Crystal Hill, Agate Creek, and Black Soil Creek. It is thought that each locality has a somewhat distinctive style of agate but the agates are mainly concentric banded amygdules, often multi-colored, often with a pistachio green color to them. They are usually banded in straight, curved, or irregular patterns. There are also seam, tube, moss, dendritic, and banded onyx types as well. Secondary minerals include amethyst, aragonite, and calcite.

In addition to the volcanic agate found here, Morrison et al. describe agate found associated with epithermal rhyolite; where hydrothermal chalcedony is indicated by boiling textures such as a feathery texture. Further, the quartz found in the area is classified by Morrison et al., as chalcedony that is banded, consisting of crustiform, colloform, and cockade types.

Thundereggs are also in the rhyolitic lava of the Thunder Egg Rhyolite member. The silica source for the thundereggs is from the ashy or tuffaceous deposits of the Big Surprise tuff. The thundereggs in the rhyolite may contain infillings of red-brown jasper.

7.16.1 Madina Agate

Agate is also found in the Western Australia province in the Pilbara craton. The boundaries of the Pilbara Craton are either tectonic, against younger Precambrian rocks of the Capricorn Orogen, or are concealed by unconformably overlying Precambrian and Phanerozoic rocks. The Fortescue Group rocks, in the Pilbara Craton, have an older granite-greenstone component formed largely between 3.5 and 2.9 Bya, and an unconformably overlying supracrustal succession called the Mount Bruce Supergroup (Thorne & Trendall, 2001). The three major and regionally concordant subdivisions of the Mount Bruce Supergroup, the Fortescue Group, Hamersley Group, and Turee Creek Group, formed between about 2.8 and 2.4 Bya. The Fortescue Group consists of upper sediments of fine clastics over a sequence of upper lavas. Middle to upper parts of the Fortescue Group are dominated by subaerial basaltic flows (the Kylena and Maddina Formations) and coastal and nearshore-shelf sedimentary and volcaniclastic rocks (Tumbiana Formation) in the north Pilbara, and by subaqueous basaltic to komatiitic lavas and volcaniclastic rocks (Boongal, Pyradie, and Bunjinah Formations) in the south. The Jeerinah Formation, the uppermost unit within the Fortescue Group, consists largely of argillaceous rocks in the north, whereas basaltic lava and volcaniclastic rocks are abundant in the south.

Agates are found as amygdules in the vesicular Maddina basalt. The lava flows are 600–800 thick and are a massive to vesicular, locally pillowed basalt, with intercalations of subordinate stromatolitic carbonate and quartz sandstone. Flow tops are silicified and the vesicles are filled with quartz (chalcedony, agate), carbonates, and chlorite.

Of particular interest is that the agate and the host rock have been subjected to low-grade metamorphism, with the rocks exhibiting the prehnite-pumpellyite metamorphic facies mineralogy and aphyric (two-phase) texture. The crystallization, banding, and fibrosity of the agates were affected. Depending on the amount or stage of metamorphism, the agates ultimately recrystallize to granular quartz with interpenetrating silica and exhibit a jigsaw-puzzle texture. Morganite quartz content is reduced to a minor constituent (Fig. 7.56).

Fig. 7.56 Agate Creek geological map (Queensland Government, Australia, 1995)

7.17 Australia Sedimentary Opal

Opal is found in Cretaceous sediments in the Great Artesian Basin in eastern Australia and is the classic example of supergene continental weathering of sedimentary and volcaniclastic rocks, with the dissolving of primary minerals and the liberation of silica, forming Opal A (Fig. 7.57).

In the Cretaceous period, 125–95 Mya, the Great Artesian Basin was a foreland basin consisting of an epicontinental sea on the Australian plate, that was moving eastward over a west-dipping subduction zone that was associated with a Cordilleran-type orogenic belt built along the Pacific margin of Gondwana (Fig. 7.58). Volcaniclastic sediments consisting of arkosic feldspars, volcanic clasts, and volcanic ash eroded from the Cordilleran volcanic belt (the Whitsunday Volcanic province of northeast Queensland Province) and were deposited in the basin. 100–65 Mya, uplift, erosion, and intense weathering of the basinal rocks occurred, removing up to 3 km of sedimentary rocks. The intense weathering produced extensive silicification within the Tertiary regolith, in which the Lower Cretaceous sediments, along with the Paleozoic and older rocks near the margins of the basin, underwent a deep, intense weathering, through a rare, acidic, oxidative weathering process, with the subsequent argillic alteration of the sediments to kaolinite and the release of silica (Dutkiewicz et al., 2015). Petrographic evidence of the silicification consisted of reworked feldspar grains that

Fig. 7.57 Great Artesian Basin and major opal mining locations (Dutkiewicz et al., 2015)

were completely dissolved and replaced by opal, with the mafic grains and detrital quartz remaining unaltered. The subsequent mobilization and precipitation of silica formed opal and the iron oxide mineral, goethite.

Trace element analysis of the opals by Dutkiewicz et al. showed that Ba was the most diagnostic element in determining the provenance of the silica. Ba content that is <110 ppm indicates silica fluids derived from a volcanic source and Ba content

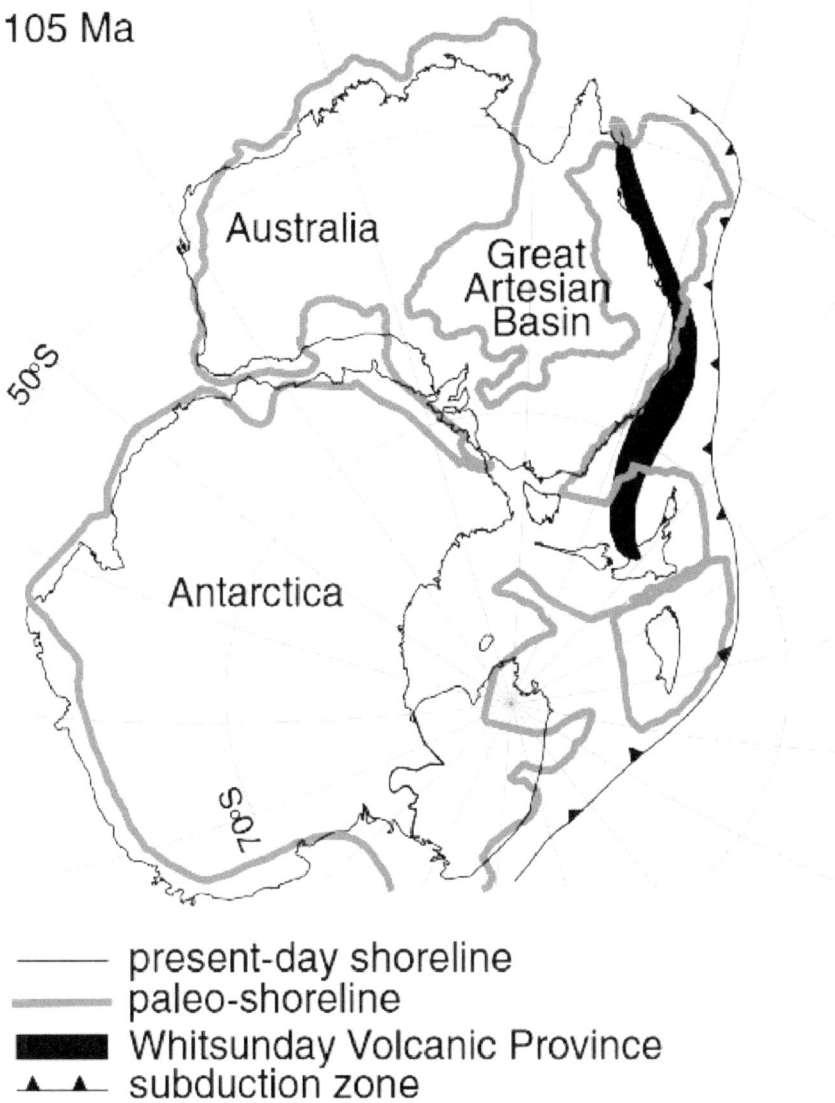

Fig. 7.58 The Australian plate 105 Mya and relationship between Whitsunday silicic large igneous province and the Great Artesian Basin (Dutkiewicz et al., 2015)

>110 ppm, indicated silica that is sedimentary in origin. The more weathered the rocks were, the higher the Ba content and enrichment. The source of the Ba is the Whitsunday Volcanic Province, which supplied volcanogenic material into the Great Artesian Basin. A small amount of the opal, however, is probably volcanic in origin. This is corroborated by the Quilpie and Lightning Ridge opal locations that have the lowest concentration of Ba and are closest to the Whitsunday Province and the Coober Pedy opal location, with higher concentrations of Ba, further from the Whitsunday Province.

From Oxygen isotope and K–Ar dating, it is thought that weathering occurred during a few discrete episodes from the Late Cretaceous period to the Cenozoic, from 130 Mya to 1,000 years B.P. Radiocarbon dating from carbon inclusions in opal, provided dates at <46 thousand years, suggesting that the formation of opal is still occurring from the Cenozoic to the present (Rey, 2013).

In the basin, silica percolated through the overlying Winton Formation, a predominantly coarse volcaniclastic sandstone, and was deposited on impermeable clay lenses within the Winton sandstone and at the contact between the sandstone and the underlying impermeable claystone of the Bulldog Shale/Wallumbilla Formation (Fig. 7.59). Additionally, opal is found in the porous and permeable zones within the sandstone, such as paleochannels, fault zones, bedding planes and as replacements of iron concretions and fossils. It is found as veins, seams, nodules, and coatings. Gentle warping of the sediments occurred ~24 Mya, with many of the opal deposits found in the topographic highs of folds.

7.17.1 Andamooka Field

The Andamooka field was found in 1930 with the opal found in the Bulldog shale. It is found in a highly weathered sandy clay called "kopi." The sandy clay grades into a conglomerate that contains pre-Mesozoic age Arcoona formation quartzite clasts. The sandy clay is underlain by the Bulldog claystone; "mud." Opal is found at the conglomerate/claystone boundary interface.

7.17.2 Coober Pedy

The field was found in 1915 in the early Cretaceous Bulldog shale. A highly weathered sandstone, silty or sandy claystone, that overlies the Bulldog shale, is bleached white by the argillic alteration of the sandstone feldspars to kaolinite. The sandstone is underlain by claystone or "mud" of the early Cretaceous marine, kaolinite, Bulldog Formation. The opal is found at the contact between the sandstone and the claystone but is also found throughout the sandstone unit in veins, bedding planes, fossils, and joints.

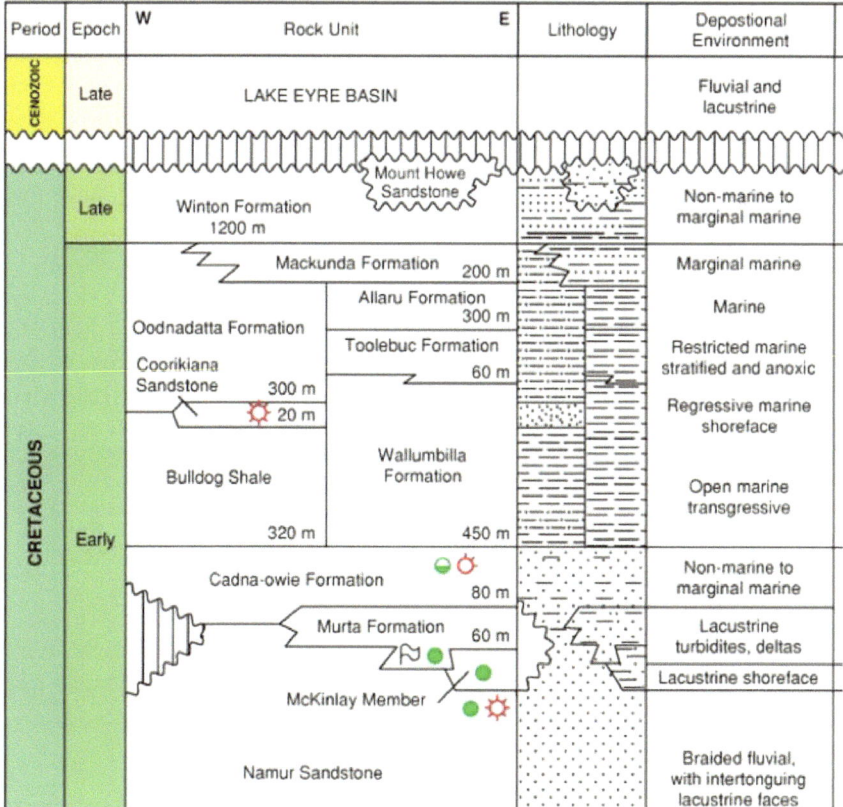

Fig. 7.59 Stratigraphic chart for the Great Artisan Basin (after Hill et al., 2015)

7.17.3   Mintabie

The field was found in 1921–22 in Early Cretaceous sediments that on-lap an Ordovi-cian, kaolinitic, fluvio-deltaic sandstone, that has cross-bedded claystone interbeds. The weathered horizon, that is an unconformity, extends to the Ordovician sandstone, which has also been weathered, making this the only field with opals from Paleozoic rocks.

7.17.4   Lambina

At this locality, opal replaces snail, belemnites, and bivalve fossils. It also occurs in veins, fissures, and cracks within the host rocks. It especially occurs in ancient pale-ochannels that have exceptional porosity and permeability, forming a good conduit for siliceous fluids. The paleochannels cut into the early Cretaceous Bulldog shale. The opal occurs in topographically high mesas. The paleochannels, where present, were originally formed in topographic lows but are now in topographic highs due to a reversal in topography, resulting from erosion of the softer surrounding weathered Tertiary and Cretaceous sediments.

7.17.5 New South Wales Lightning Ridge Area

Found in the 1880s in the Wallumbilla sandstone member that overlies the Bulldog claystone. A silicified zone, mms to one meter thick, is at the sandstone/claystone boundary. Opal is found in the silicified zone; also, on claystone lenses (Late Albian, kaolinite-rich, Finch Clay flood-plain facies member) within the sandstone. It is also found in paleochannels as nodules, seams, iron concretion and fossil replacements.

7.17.6 Winton/Quilpie

Opal is found in the late Albian to Cenomanian fluvial to volcaniclastic Winton Formation, consisting of red-brown ironstone, goethite, kaolinite, with minor quartz and illite.

7.17.7 New South Wales, White Cliffs

Opal is found in the conglomeratic Doncaster member of the Wallumbilla Formation that overlies the Bulldog shale. It is found as seams and as coatings on Devonian quartzite boulder intraclasts in the basal conglomerate.

7.18 Argentina

Condor agate was discovered in 1993 in Valle Grande, 30 km southwest of San Rafael City, in Mendoza Province. The exact locations are somewhat nebulous and when an exact location is found, it is hampered by the lack of any geological description of the area. The Condor agate, however, is reasonably accurate as to its location and geology, but even there, questions arise as to its accuracy. It is described as found in decomposed rhyolite near Hills of Valle Grande, near San Rafael City in rocks that are 30 Ma old. However, there are no rhyolitic rocks in the area, but instead we find that it is the location of the Payenia Volcanic field (Fig. 7.60). These lava fields lie in the back-arc region of the Andean Volcanic belt, which was formed by the subduction of the Nazca or Pacific plate beneath the South American plate. Ramos and Folguera (2011) described the rocks as products of volcanism that commenced during Pleistocene times, 168–82k years ago, and formed the Liancanelo basin, Nevado basin, Salado basin, Pampas Ondulados, and Payun Matra lava fields. The volcanism consisted of the Payun Matri shield volcano, lava flows, and scoria cones. The lava flows generally exhibit pahoehoe structure and are considered alkaline flood basalts. If the Condor agate is found in these rocks, it would appear that the Condor agate would be the youngest agate found to date. The agates are of a deep, ruby, red, black, white, lavender, and occasionally yellowish-orange color (Fig. 7.61). The colors reflect the weathering and alteration minerals hematite/goethite, and magnesite permeating the agate sol. The Puma agate, also found in the area in probable Paleozoic limestones, is a sedimentary type of agate that contains fossiliferous coral limestone and reflects deposition and formation in a limestone deposition environment.

   Another major type of agate is found in Las Plumas, Chubut province. The geology of the area is described by Manassero et al. (2000), as a silicic volcanic intraplate association related to crustal extension of the Gondwana Continent rifting. The rocks are Jurassic stratified tuffs and volcaniclastics, with the source of the tuffs originating from rhyolitic and dacitic volcanoes 350 km west of Las Plumes. The volcaniclastics and tuffs are underlain by Middle Jurassic ignimbrites; which sets up the geologic stage or picture for the occurrence of thundereggs or agatized geodes, rather than the vesicular-filled agate found in basalts or andesites. Interestingly, it exhibits little or no banding and is almost a massive type of chalcedony. Many of the geodes have botryoidal textures, reflecting the colloidal and surficial nature of the agates (Fig. 7.62).

7.19 China

7.19.1 Xuanhua Agate

The area of this deposit is about 20 km^2, is located in Hebei Province, and is situated on the northern margin of the North China Craton. With the subduction Paleo-Pacific plate underneath the Eurasian plate in the Late Jurassic to Early Cretaceous period, a series of intermediate-acid igneous rocks erupted. The stratigraphy of this agate area consists, from bottom to top, of the Jurassic Tiaojishan formation, the Jurassic Houcheng formation, the Cretaceous Zhangjiakou formation, and capped by Cenozoic Quaternary loess (Fig. 7.63). The agate, known as "Xuanhua" agate occurs amygdaloidal, or as veins in the vesicles and fractures in a trachyte lava flow, in the Jurassic Tiaojishan formation. The agate displays an intense red, yellow, and orange colors (Zhou et al., 2021).

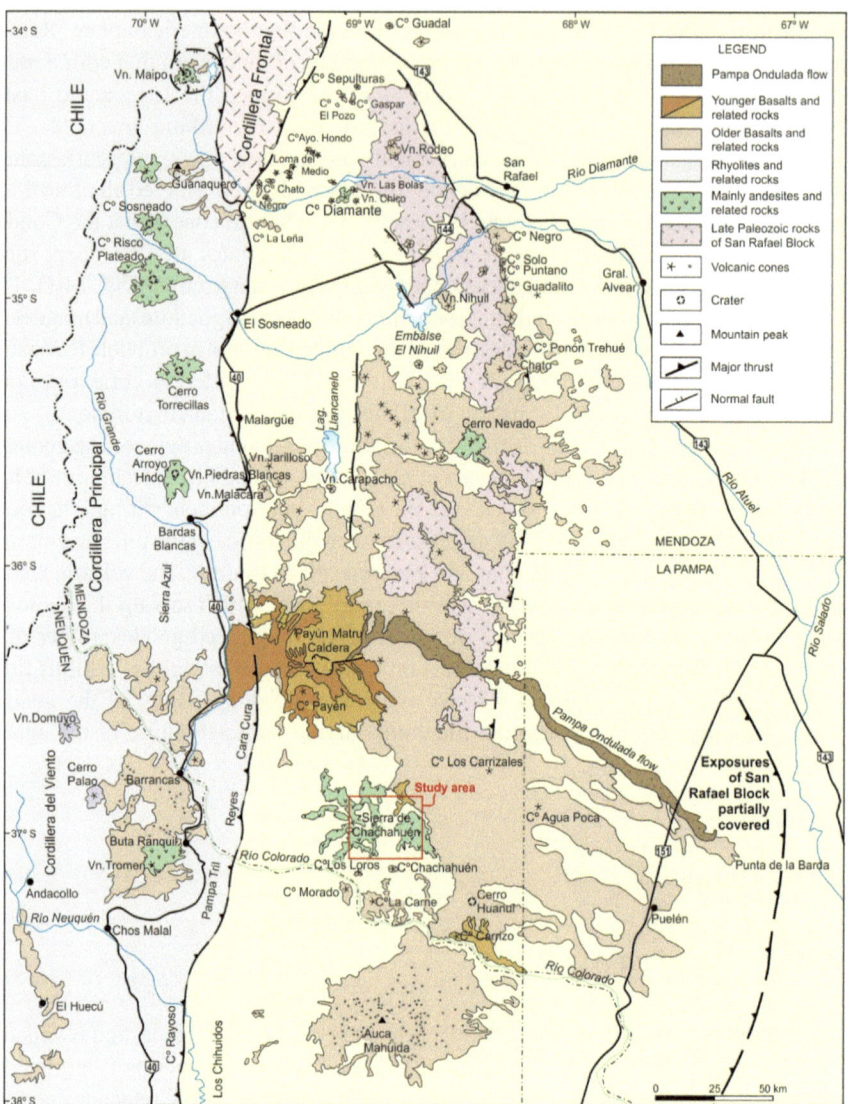

Fig. 7.60 Volcanic setting for the Condor agate, Mendoza Province, Argentina (Ramos & Folguera, 2011)

7.19.2   Nanhong Agate

A rift system developed prior to the main stage of flood basalt eruptions of the Middle-Late Permian Emeishan large igneous province (Figs. 7.64 and 7.65). The rifting progressed in pulses, with an initial phase of normal faulting followed by rapid deposition of breccias and clastic sedimentary deposits. Later there was

Fig. 7.61 Condor agate

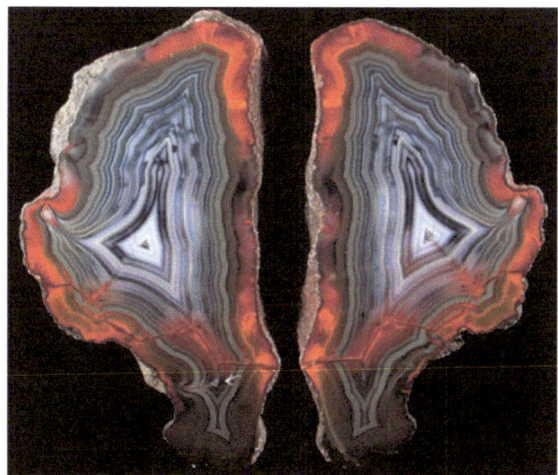

Fig. 7.62 Geode-type crater agate with botryoidal texture

lower-energy deposition of sandstone, mudstones and dolomites with accompanying hydromagmatic deposits and rhyolitic eruptions (Shellnut, 2013).

The "Nanhong" agate was discovered approximately 10 years ago in Meigu County, Sichuan Province (Xu et al., 2021). The agate is found in the vesicles and fissures of a Permian Emeishan flood basalt lava flow in the Permian Leping Formation. The agate is colored a deep red or is colorless with red hematitic inclusions.

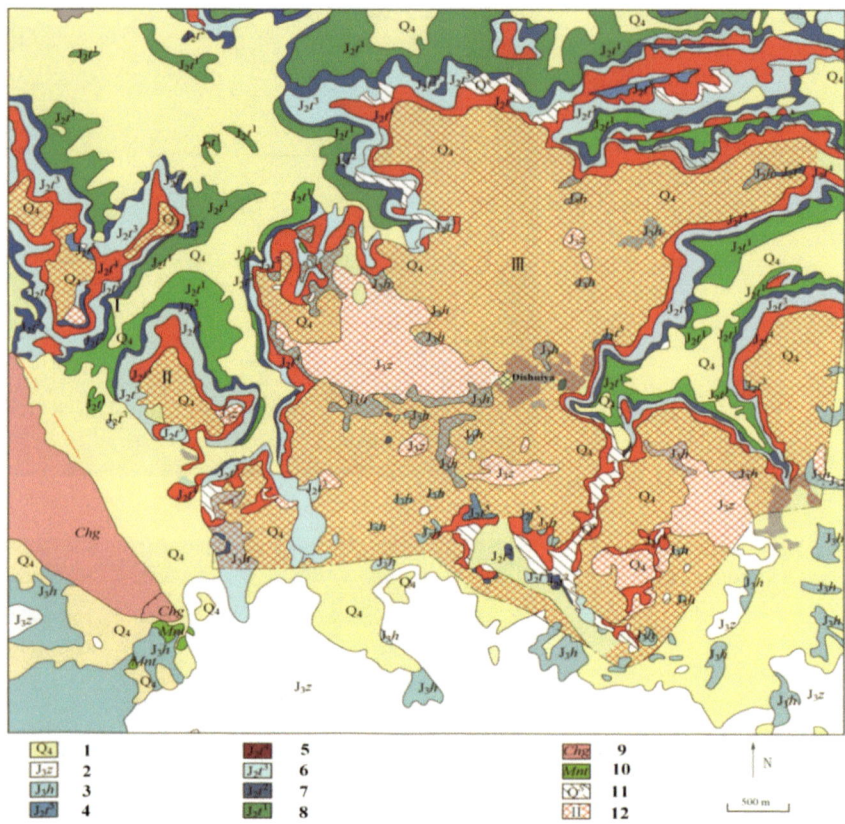

Fig. 7.63 Geological map of Xuanhua agate deposit and its adjacent areas: (1) quaternary loess; (2) rhyolitic tuff; (3) siltstone; (4) agglomerate tuff and tuffaceous breccias; (5) vesicular trachyte; (6) massive trachyte; (7) tuffaceous siltstone and volcanic breccias; (8) basalt; (9) dolomite of the Mesoproterozoic Changcheng formation; (10) bentonite deposit; (11) Qml; (12) agate-bearing zones and their number (Zhou et al., 2021)

7.19.3 Zhanguohong Agate

The agate deposit is located near Beipiao, Liaoning province, China and is situated adjacent to the northeastern part of the Jurassic Beipiao basin in the east Yanshan belt, northern North China Craton. The NEE striking Lingyuan–Beipiao–Shahe fault zones and the NNE striking Chaoyang–Yaowangmiao fault zones intersect in this area. Two major rock units are present in the Beipiao district: a Cretaceous sedimentary and volcanic unit and a pre-Cretaceous unit (Fig. 7.66). The latter is comprised of Archean crystalline basement rocks and Jurassic clastics and volcanics. The Cretaceous sedimentary and volcanic unit includes, from bottom to top, the Early Cretaceous Yixian, Jiufotang, Fuxin formations, Late Cretaceous Sunjiawan Formation, and the Tuchengzi Formation; the Tuchengzi formation separated from the Sunjiawan

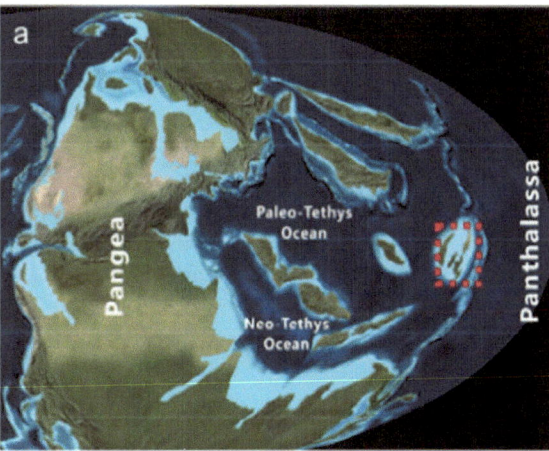

Fig. 7.64 Position of the Emeishan flood basalts in Middle Permian (Wang et al., 2022)

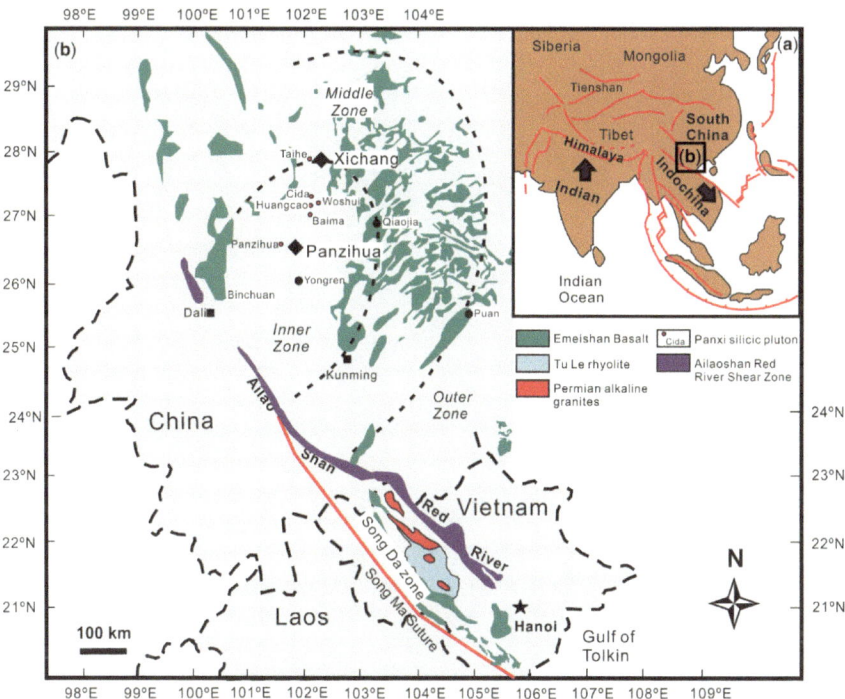

Fig. 7.65 Location of the Nanhong agate in the Permian Emeishan flood basalt province and in the Sichuan Province northeast of Panzihua (Shellnutt, 2013)

formation by an unconformity. The overlying Yixian Formation consists primarily of basalt in the lower part, and andesite, intercalated with volcaniclastic rocks, with minor conglomerate, sandstone, and shale interlayers, in the upper part (Zhang, 2020).

The Zhanguohong agate deposits locally occur in the volcanic breccias of the Yixian Formation Early Cretaceous in age (135–120 Ma). The Lower Cretaceous strata were formed in rift basins due to the intense lithospheric extension and thinning in eastern China, and throughout east Asia, in the Late Mesozoic. With prolonged lithospheric extension, the Cretaceous intermediate, felsic, and alkaline volcanics rocks were emplaced.

The agate generally shows massive and banded structures, with red, yellow, and/ or white layers or zones.

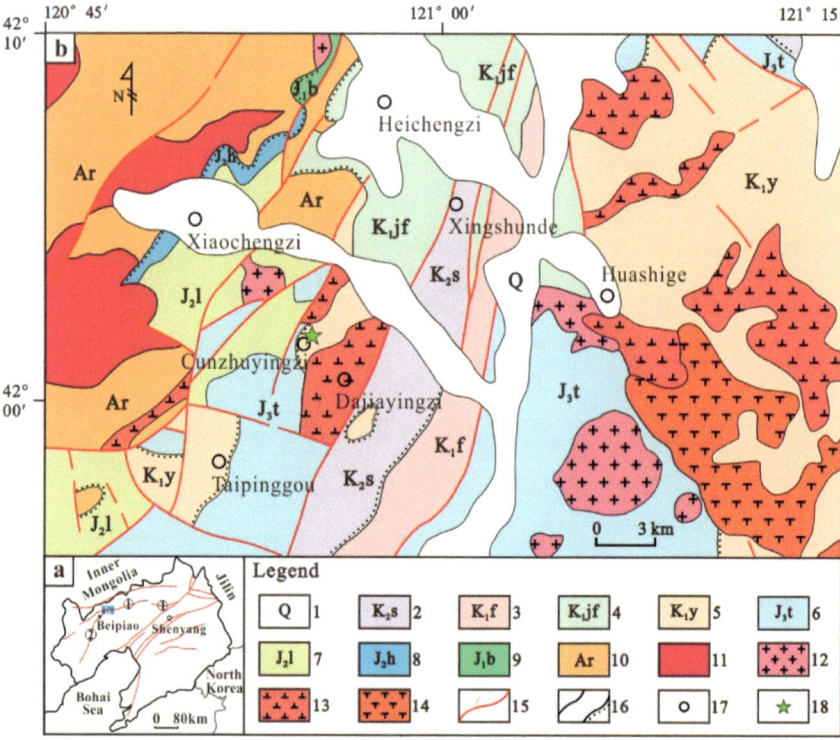

Fig. 7.66 Geological Map of the Zhanguohong Agate Location, Liaoning Province. (1) Quaternary sediments; (2–5) Cretaceous sedimentary–volcanic rocks (2—Sunjiawan Fm; 3—Fuxin Fm; 4—Jiufotang Fm; 5—Yixian Fm); (6–9) Jurassic sedimentary–volcanic rocks (6—Tuchengzi Fm; 7—Lanqi Fm; 8—Haifanggou Fm; 9—Beipiao Fm); (10) Archean basement rocks; (11) Permian–Jurassic intrusive rocks; (12–14) Cretaceous intrusive rocks (12—granitoid rocks; 13—diorite porphyrite; 14—syenite porphyry); (15) fault; (16) conformity and unconformity; (17) toponym; (18) sample location. O1 Lingyuan–Beipiao–Shahe fault; O2 Chaoyang–Yaowangmiao fault. Fm: Formation (Zhang et al., 2020)

7.19.4   *Yuhuatai Alluvial Agate*

Agate nodules, of a type similar to the Montana moss agate and the Indian black-skin agate, are found in the Yuhuatai alluvial gravels of the ancestral Yangtze River and its tributaries, the Quinhuai and Chuhe Rivers. Like the Montana moss agate, the original host rocks have been eroded away. The best localities are near Nanjing and Jiangsu.

7.20   Montana Moss Agate

The Montana moss agate is found in the found in alluvial gravels (Flaxville gravels) of the Yellowstone River, north of Yellowstone Park to the North Dakota border and its tributaries and terraces. The best collecting areas are between Sydney and Custer, and between Glendive and Forsyth.

The host rock, basalts, or andesites have been eroded away and nothing remains of the source and host rock (Fig. 7.67). The central locality and source area of the agate is thought to be the Yellowstone Park area, where the Absaroka Volcanic Field is located, with the Yellowstone River eroding the lava flow, transporting and distributing the agates along the Yellowstone sediments. The Absaroka volcanic field consists of Eocene-aged deeply eroded andesitic and basaltic stratovolcanoes and andesitic, basaltic, and dacitic volcaniclastics (Smedes & Prostka, 1972 and Fig. 7.68). In trying to determine the nature of the host rock, is there a possibility of some host rock fragments that would still adhere to the agate and thereby determine the identity of the host rock?

The agates have black, pyrolusite (manganese oxide) and reddish-brown hematite or limonite (iron oxide) dendrites in a yellowish chalcedony with white banding. The outer surface is a drab brown or black color. Frequently, the agates have a white crust derived from sun bleaching. The agates resemble the Indian black- or yellow skin agates (Fig. 7.68).

"The River Runs North: A Story of Montana Moss Agate" by Tom Harmon contains various stories and anecdotes about the agate. Of note and interest, a biography of George Armstrong Custer, "Son of the Morning Star" by Evan S. Connell notes that the Little Big Horn River, a tributary of the Yellowstone River and site of Custer's Last Stand, was frequented by various contemporaries searching for Montana moss agate.

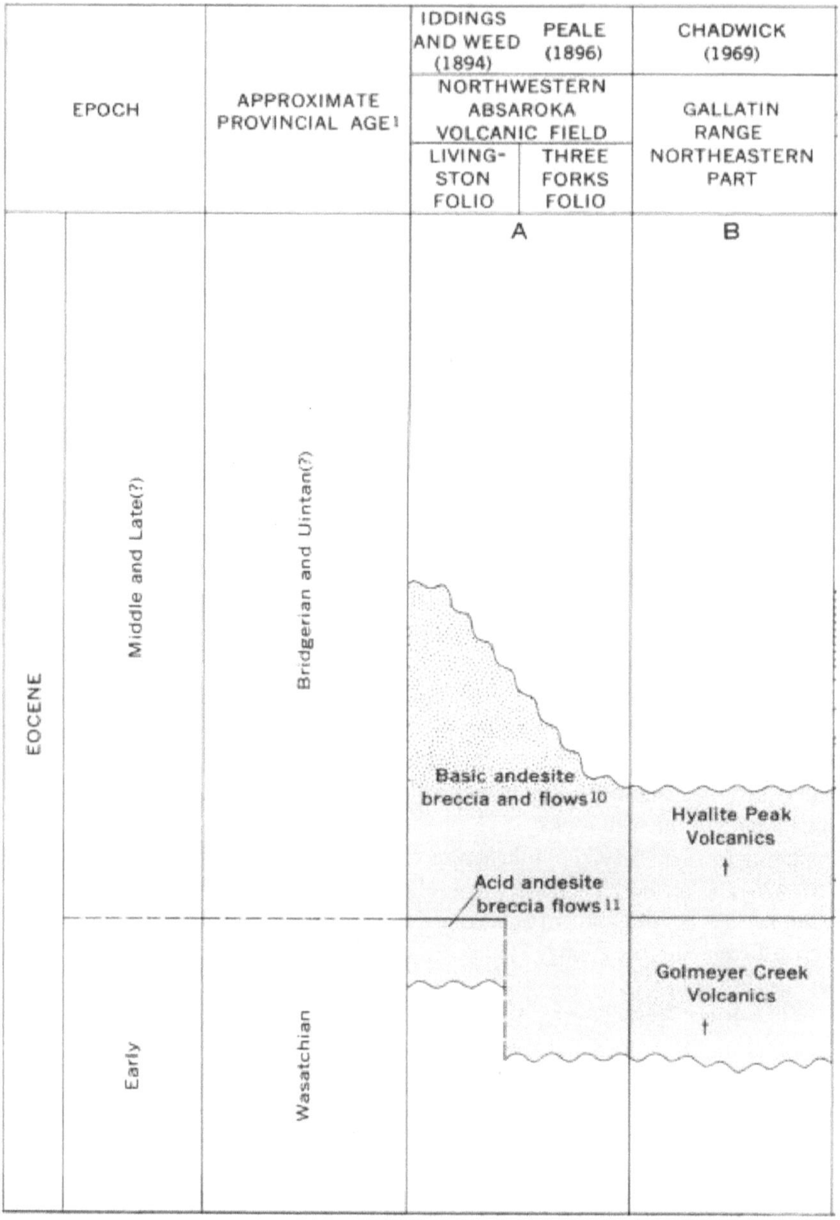
EPOCH
APPROXIMATE PROVINCIAL AGE1
IDDINGS AND WEED (1894)
PEALE (1896)
CHADWICK (1969)
NORTHWESTERN ABSAROKA VOLCANIC FIELD
LIVINGSTON FOLIO
THREE FORKS FOLIO
GALLATIN RANGE NORTHEASTERN PART
A
B
EOCENE
Middle and Late(?)
Early
Bridgerian and Uintan(?)
Wasatchian
Basic andesite breccia and flows10
Acid andesite breccia flows11
Hyalite Peak Volcanics
Golmeyer Creek Volcanics

Fig. 7.67 Absaroka volcanic field stratigraphic chart indicating erosion of andesite flows in Yellowstone National Park (Smedes & Prostka, 1972)

Fig. 7.68 Montana moss agate

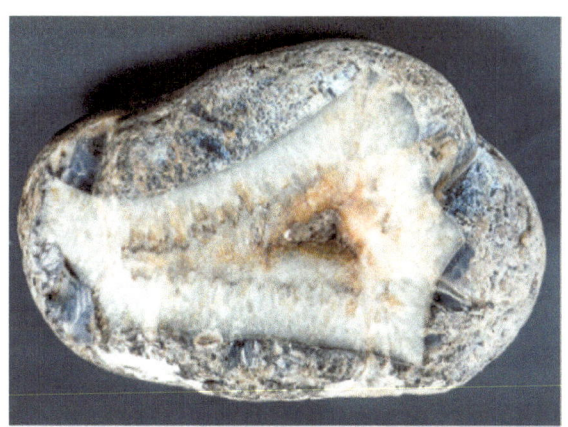

Chapter 8
Sedimentary Agate

We have concluded the examination of the major volcanic agate deposits of the world and now turn our attention to sedimentary agates. The author is aware of sedimentary agates in the United States of America (which we will describe), but there must be sedimentary agates elsewhere. We will start our examination of sedimentary agates in Kentucky, where it is the state gem stone, and work our way west, as there is a distinct possibility that these agates are related in their formation and occurrence.

8.1 Kentucky

The Kentucky Agate district, where the agates are found, is comprised of six counties: Powell, Madison, Estill, Lee, Rockcastle, and Jackson; the southerly area otherwise known as the "Knobs" Region. The Knobs Region of Kentucky is a wide "u"-shaped arc that extends for about 230 miles through central Kentucky. The eastern portion of the Knobs Region is bounded on the east by the Cumberland Plateau. The region is made up of a complex system of creeks and rivers that have eroded the edge of the Cumberland Plateau into a network of ridges and conical hills called "knobs." The agates formed in a narrow band in the Nada member of the Borden Formation located near the contact with the overlying dolomitic Renfro member (Fig. 8.1). The host rock is an olive green to gray siltstone. Most of the agates are found as alluvium in the stream beds.

The Borden Formation is part of the Appalachian Foreland Basin, that was filled with fluvial, deltaic complex sediments, derived from the ancestral Appalachian orogenic belt (now termed the Neo-Acadian orogenic belt) formed during early- to middle-Mississippian times (Ettensohn et al., 2012). The stratigraphic succession grades from deeper water fan-type environments in the Nancy–Cowbell members up through the delta destruction facies of the Nada, then onward to a carbonate facies where terrestrial influence is all but absent.

A. Zarins, *The Geology of Agate Deposits*,
SpringerBriefs in Earth System Sciences, https://doi.org/10.1007/978-3-031-67929-2_8

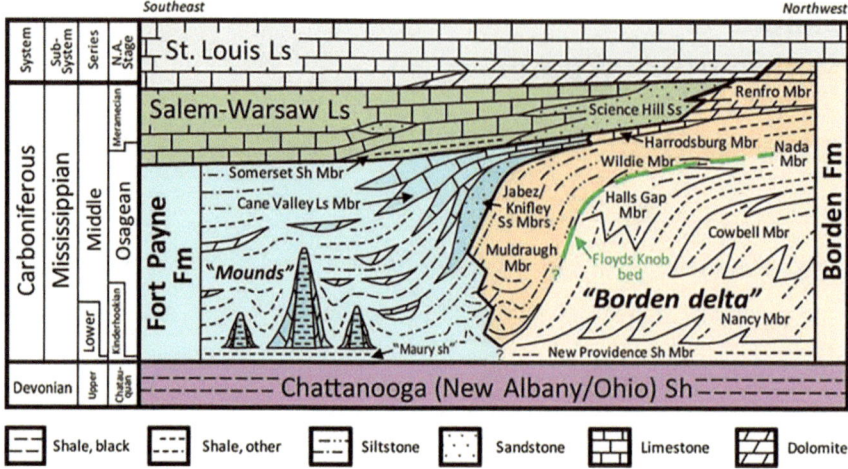

Fig. 8.1 Mississippian stratigraphy cross section across Kentucky (Greb et al., 2011)

The Borden Formation is comprised of siltstones, shales, and dolomitic limestones in the agate-producing zone of Kentucky, consisting of the Cowbell, Nada, and Renfro members. Agates can be found in all three of these intervals but are primarily in the top of the Cowbell and in the Nada members. Geodes in high concentrations are found in the overlying Muldraugh member as well.

Voids in the sediments formed from the dissolution of organic matter and fossils that were later filled with evaporitic minerals such as anhydrite, gypsum, barite, and calcite/aragonite (Greb, 2012). Pseudomorphs of these fossils are frequently found in the agates.

The overlying Renfro and Muldraugh members consist of argillaceous, cherty, dolomitic limestone and a cherty, dolomitic calcareous siltstone, respectively (Udgata & Ettensohn, 2020). These delto-fluvaic rock types indicated a sub-areal, arid, alkaline, high pH, sabkha-type environment that formed evaporitic minerals in the voids from leaching ground water. The silica source is derived from the contemporaneous and subsequent weathering of these rocks creating an amorphous siliceous ooze derived from the dissolution of silicious fossils such as diatoms, radiolarians, and sponge spicules.* The voids were filled and formed the agates; agates indicating a significant unconformity. The red bands of the agates consist of hematite, a weathering-derived iron-oxide mineral, that is an impurity in the chalcedony, forming the bands along crystallization fronts (Fig. 8.2). The rhythmic banding suggests multiple weathering and erosional events. Morganite is found in the agates, as well, in a high weight percentage, replacing evaporitic and biogenic carbonate minerals, and indicating an arid evaporitic environment.

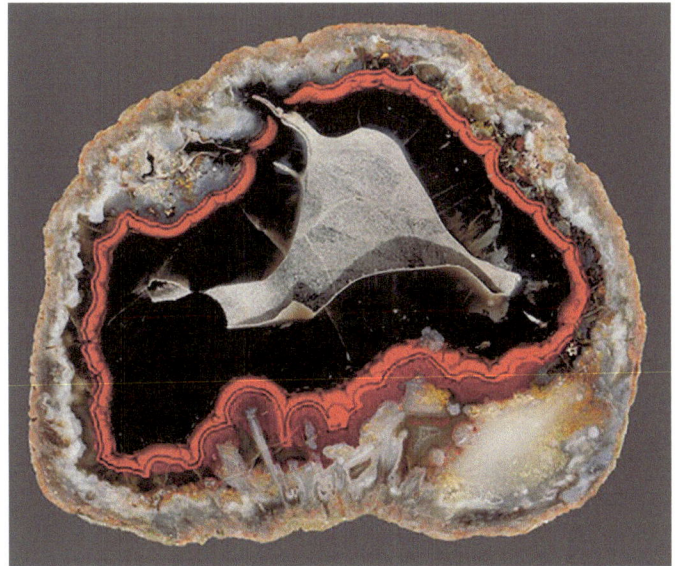

Fig. 8.2 Kentucky agate

*An alternate theory for the source of the silica posits that silica-enriched fluids migrated through faults such as the Irvine-Paint Creek Fault in the knobs area, inundating the area with an abundance of SiO_2. There are many places where silica is still in fractures all throughout the top of the Cowbell.

8.2   South Dakota Fairburn (Teepee Canyon) Agate

The Fairburn agate is found near the towns of Fairburn and Interior, South Dakota, south of Kodoka, and in the White River Badlands, in the surficial loose, late Cenozoic-aged gravels east of the Black Hills. It is now generally accepted and proven that the famous "Fairburn" agates actually originated from the Upper Pennsylvanian-aged Minnelusa Formation and the Lower Mississippian Pahasapa Limestone in the Black Hills of South Dakota, west of Custer. The Minnelusa according to Fagnan (2009) and Darton (1901) is mainly a buff and reddish, fine-grained, massively bedded sandstone that contains reddish or grayish thinly bedded limestones and sandy shales with occasional cherty layers. At the agate localities, the middle portion of the formation consists of a yellowish gray 30 ft-thick sandy limestone at the top, underlain by a tan, silicified, sandstone interbedded with shale and at bottom, a silicified gravel horizon. At the top of this unit, containing silicified fossils of *Chaetetes milliporaceus*, a prominent marker bed is found. The lower portions of the Minnelusa consist of yellowish gray thin-bedded limestones that contain red and white chert nodules and thin, interbedded sandstone and shale layers, underlain by a 40 ft section of tan or red, medium-to-coarse grained, cross-bedded sandstone, and siltstone. A prominent disconformity separates the Minnelusa from the underlying Lower Mississippian Pahasapa limestone. The Pahasapa limestone is a gray to light tan cavernous limestone (Wind Cave and Crystal Cave are prominent caves found within this limestone) and dolomitic limestone that is massively bedded in the upper portion and thick bedded dolomitic, sandy limestone in the lower portion. It forms prominent cliffs and exposures and contains rugose coral and spiriferid brachiopod fossils. It is 430 ft thick. Darton describes the upper part of the formation as being siliceous and flinty and stained red. At the very top, a thin red shaley bed contains oval concretions of hard silica, from 6″ to 2 ft in diameter. The fossils *Spirifer rockymontanus*, *Seminila dawsoni*, *Productus*, and *Zaphrentis* are found, indicating a lower Carboniferous age.

Early diagenesis of the sediments occurred with the formation of voids, with many of the voids created from decaying organics. The organic matter produced anaerobic conditions and a high pH environment. The Pahasapa Limestone at the disconformity surface is a weathered horizon indicating an arid, evaporitic environment. Evaporitic minerals filled the voids with later silicification replacing the evaporitic minerals. Silicification took place prior to compaction and lithification of the sediments, the most likely source of silica solutions being the abundant sponge spicules which characterize the peritidal dolostones, and abundant volcanic ash and "Fullers Earth" found in the area.

The agate is found in the sandstone, siltstone, limestone, and shale of the lower basal portions of the Minnelusa Formation and in the shales and limestones beneath the disconformity in the upper portions of the Pahasapa Limestone at Teepee Canyon and other Black Hills area locations. In early literature, agates from this locality were sometimes referred to as "Hell's Canyon agates." Teepee Canyon is located on the north side of Highway 16, about 18 miles west of the town of Custer, South Dakota. Two main areas (West Teepee Canyon and Sawmill Spring [FS Road#456] area) about 1–2 miles west of the Jewel Cave monument boundary are where the main diggings are located (Fig. 8.3). Agates occur as small to large, tan to chocolate-colored chert nodules. Most nodules are plain and devoid of fortification agate, or have plain red jasper. It is said that any nodule containing more than 25–30% agate is rare. The gem-quality agates display fine fortification banding with alternating layers of red, black, orange, yellow, pink, cream, and white, and, in most specimens, surrounded by a tan-colored siltstone (Figs. 8.4 and 8.5). The centers of the agates typically contain calcite (common) or drusy quartz and (more rarely) amethyst. The finest specimens display the trademark "holly leaf" fortification pattern. Away from the traditional collecting areas, the agates show duller tones and banding. Of interest and possible significance, Clark (2005), noted that Fairburn agates lack certain features common to volcanic agate, viz., eyes, tubes, horizontal banding (Uruguay structure), pseudomorphs, and sagenitic structure.

Fig. 8.3 Minnelusa Formation, Jewel Cave National Monument, South Dakota

Fig. 8.4 Thin section of a Fairburn Agate; 27 × 46 mm, ×2, polarized light

8.3 Montana Dryhead Agate

Dryhead agates are found in a red shale bed along Dryhead Creek close to the headwaters of the creek at the Big Horn River. The red shale bed belongs to the Early Permian Phosphoria Formation (the Embar Formation of earlier reports). In Permian times, the area consisted of a shifting ocean shoreline along the Western

Fig. 8.5 Fairburn agate

portion of the North American continent. There is a sliver of the Permian Rocks in the Dryhead Creek area. It is separated from the underlying Ten Sleep sandstone by significant erosional unconformity, representing a period of considerable weathering and erosion. The Triassic Chugwater sandstone overlies the Phosphoria rocks, probably by another erosional unconformity, though it is not clear if this unconformity exists. The agate-bearing rocks are situated between the East Pryor Mountains on the west and the Bighorn mountains on the east. According to Richards (1955), the rocks of the Phosphoria Formation, in the agate area, consist of 25–40 ft of white, siliceous sandstone, underlain by 70–90 ft of red and yellow shales. To the south, the rocks grade into a maroon sandstone and siltstone with a basal chert conglomerate. It is underlain by a significant unconformity. The sandstones of the Pennsylvanian Ten Sleep and Amsden Formations and the Mississippian karsitic limestones of the Madison Formation underlie the Phosphoria Formation and are thought to be the equivalent formations of the Minnelusa and Pahusapa limestone, respectively, in the Black Hills area. Interestingly, the upper part of the Ten Sleep Sandstone, beneath the unconformity, contains cavities lined with drusy quartz, indicating the presence of circulatory siliceous waters (Thom, 1935).

Seemingly, the Fairburn agate, found in the Black Hills area of South Dakota, is identical in appearance to the Dryhead agate that also formed on an unconformity in the Lower Mississippian to Upper Pennsylvanian time. Since agates are a surface or near-surface phenomena that make them good indicators of unconformities, raising the possibility that one is dealing with the same or identical unconformity and time transgressive/regressive units. Moreover, the agate, itself, is found in a gray siltstone

Fig. 8.6 Dryhead agate

matrix, as is the Fairburn agate (Fig. 8.6). In addition, both agate types are almost identical in appearance, having the same red banding.

The origins of the silica for the agates are probably biogenic, as attested to by the fossiliferous content of chert nodules in the area. However, Götze (2009) analyzed the Dryhead agate, among other analyses, by cathodoluminescence (CL) for trace element composition and isotopic oxygen analysis for temperature determination. The analyses results were inconsistent in regards to the usual parameters of sedimentary agate formation, indicating that agate formation involved a hydrothermal component or process.*

*U detected at 70 ppm with transient blue CL and temperature of formation <120 °C (28.9 0/00 for quartz and 32.2 0/00 for chalcedony).

8.4 Keokuk, Iowa Warsaw Formation Geodes

Mississippian-aged geodes are found in three related areas of the United States: the Iowa, Illinois, and Missouri area in the Lower Warsaw and Upper Keokuk Limestone Formations; the Indiana and Kentucky Harrodsburg Limestone and Muldraugh Formations; and in the Woodbury, Tennessee Fort Payne Chert, Maury Shale Member. The Keokuk geodes are the more well-known geodes, and will be further described as the "type" geode.

D. D. Owens of the United States Geological Survey first described the geology and the geodes of the Iowa, Illinois, and Missouri tri-state area in 1852 (Fig. 8.7). The geodes are found in the Lower Mississippian (Valmeyeran) Warsaw Formation (Fig. 8.8). The Warsaw Formation consists of a calcarenite limestone (sand-sized fossil fragments in a micritic lime mud) with interbedded shale and dolomite; the dolomite replacing the micritic lime mud. It was an open, low-energy, marine environment formed beneath the wave base. The Warsaw Formation and the underlying Keokuk limestone and Burlington limestone are the lateral equivalents of the deltaic Borden Sandstone in Kentucky, where the Kentucky sedimentary agate is found (Frankie & Jacobson, 1998).

The region was part of a southward regressing epicontinental sea in Lower Mississippian times, with the geodes occurring only in argillaceous dolomites and dolomitic mudstones; never in the overlying and underlying biocalcarenites (Hayes, 1964). Further, the geodes are mainly confined to the Lower Warsaw Formation, are associated with argillaceous dolomites and dolomitic mudstones, occur in zones or beds, initially are round, are conformable with laminations in the enclosing rocks, are uniform in size in each zone or bed, and consist of a chalcedonic outer shell and an inner layer of crystals that range from quartz and calcite, to pyrite, ankerite, magnetite, hematite, kaolin, aragonite, millerite, chalcopyrite, sphalerite, limonite, smithsonite, malachite, gypsum, fluorite, barite, marcasite, goethite, and

Fig. 8.7 Location of Keokuk Geodes (Hayes, 1964)

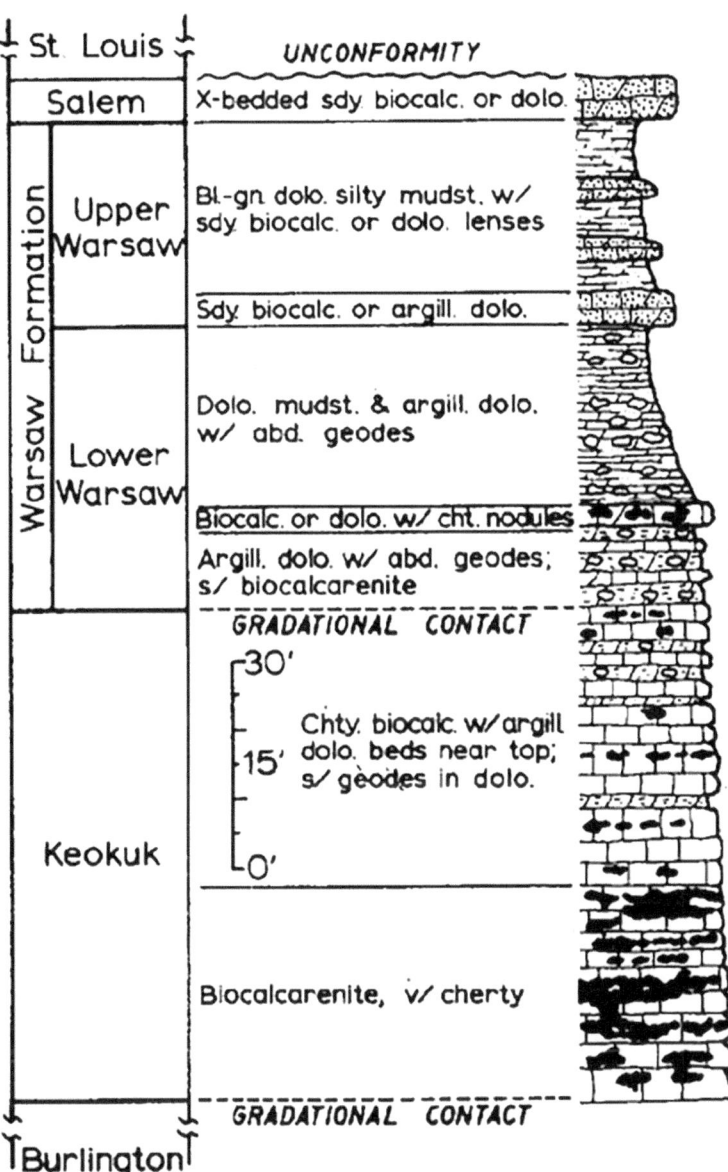

Fig. 8.8 Stratigraphic column of the lower Mississippian formations; Keokuk, Iowa Tri-state Area (Hayes, 1964)

pyrolusite. Crushed geodes, conformable laminations around the geodes, an invasion of dolomitic mudstone in the interior of some of the geodes, slickensides on exterior clay surfaces indicate that the sediment was still soft while the geodes were forming. The origin of the voids could be linked to decaying of organic matter, or the dissolution of calcitic or anhydritic concretions (Hayes, 1964; Hess, 1976; or Greb, 2012). Later silicification took place prior to compaction and lithification of the sediments, the most likely source of silica solutions being the abundant sponge spicules which characterize the associated peritidal dolostones. Compaction of the sediments occurred before the end of the Pennsylvanian period as indicated by the presence of the crushed geodes and indicate that the precipitation of silica occurred early in the sedimentation history (Hayes, 1964).

Hess, 1976, identified three different types of geodes. The first type is a silica-type geode containing crystallized quartz adjacent to the outer chalcedonic layer, with calcite composing <50% of the mass. The second type is a silica-calcite geode with recrystallized calcite; the calcite being >50% of the geode mass. The third type consists of a chalcedonic rim geode, with no silica and a brown and clear calcite in the interior; this geode is rarer and restricted to the Lower Warsaw Formation.

Mahaffray and Finkelman (2022), identified over 15 minerals in the Keokuk geodes that seem to be associated with the later stages of geode paragenesis, in a process that occurred over thousands to millions of years. They also identified the REEs chromium, molybdenum, nickel, copper, tin, zinc, and lead, seemingly reflecting the geochemistry of the Mississippi Valley lead–zinc deposits of the Tri-State district of Oklahoma, Kansas, and Missouri and the lead–zinc deposits of southeast Missouri. The minerals stages identified are: Stage 1—chalcedony, quartz, and pyrite; Stage 2—calcite crystals; Stage 3—siderite, rhodochrosite, kaolinite; Stage 4—calcite w/kaolinite, dolomite, siderite, hollandite, barite, and celestite; and Stage 5—bitumen, sylvite, and halite (Fig. 8.9).

Fig. 8.9 Keokuk Geodes with quartz, calcite, and bitumen interior

Chapter 9
Discussion and Conclusion

"Volcanic" agate deposits are chiefly found in tholeiitic, calc-alkaline igneous provinces, in basalts and andesites. Occasionally they are found in trachytes and latites. Little, visible hydrothermal alteration of the host rocks is not found at most sites, with hydrothermal deposition being implied by the deposition of zeolites and chalcedony; the zeolites forming from the leaching of the deeper host rock. On the other hand, weathered and altered lava flows and intercalated basaltic tuffs between the lava flows have altered mineralogies of hematite, limonite, goethite, kaolinitic and montmorillonitic clays and zeolites, such as clinoptilolite. The occurrence of bentonite, an alteration product of volcanic ash and a source of silica, is present at some sites, such as Chihuahua, Mexico, and Iran. The basalt, however, can be a source of silica as when the basalt is contacted by surface water; the basalt altering to agate, hematite/limonite/goethite, chlorite/celadonite, and calcite (if limestone/lime sediments are present). The flower garden agate of the Sheep Canyon basalt of West Texas being an example.

Agate formation seems to be a near-surface/surface phenomena. Ascending hydrothermal waters, probably late-stage magmatic fluids, has quartz, barite, calcite, pyrite, and fluorite initially precipitating at depths of approximately 250 m and temperatures <200 °C. At depths of 150 m to the surface and temperatures of 150–20 °C, quartz becomes chalcedonic and "agatitic," indicating deposition at low temperatures and deposition in colloidal sizes. The zeolites, especially thompsonite, calcite, and celadonite, are minerals closely associated with agates. Clays are deposited at depths of approximately 50 m to the surface, at temperatures <100 °C. Opal/cristobalite is formed at the surface at surface temperatures. Hematite, limonite, goethite, manganese (pyrolusite), and magnesium (magnesite) are deposited at all depths when present. Surface/near-surface precipitation and deposition of monomeric silicic acid and colloidal silica solutions at temperatures <100 °C occur with the mixing/commingling of surface meteoric water and groundwater. The higher the temperatures and the higher the pH, the more easily ashes and tuffs are altered and dissolved.

© The Author(s), under exclusive license to Springer Nature Switzerland AG 2025   143
A. Zarins, *The Geology of Agate Deposits*,
SpringerBriefs in Earth System Sciences, https://doi.org/10.1007/978-3-031-67929-2_9

Thundereggs can be thought of as an agate "prototype," with incomplete banding; initiation of which is probably due to temperature and pressure decreases (conversely, geodes that are initially chalcedony lined or layered are incomplete agates, due to the cessation of the flow of colloidal infilling material). The formation of thundereggs is categorically a surface phenomenon and is a horizon present in many of the agate provinces. The formation of lithophysae and the creation of spherulites in tuffs and ashes, with the formation of feldspar and quartz by intraformational hydrothermal and commingled meteoric hot waters and the subsequent expansion of steam, create the five-pointed star cavity. With subsequent weathering, a cavity is formed and filled with siliceous oozes from the dissolution of the tuffs and ashes, forming the agatized thunderegg.

Agate formation temperatures range from 20 to 230 °C as indicated by hydrogen and oxygen isotope ($\delta^{18}O$ vs. SMOW and δ^2DH vs. SMOW) studies. The strong variations in the $\delta^{18}O$ values suggest fluid mixing under non-equilibrium conditions. Exact calculations are not possible due to the unknown origins of the primary fluids. Water is present in agates as OH (Si–OH silanol groups). Fluid inclusions in water, with minimum homogenization temperature values, indicate values between <100 and >500 °C. This more or less confirms the temperature formation values of the oxygen isotope studies. Cathodoluminescence (CL) spectra of agates show at least three broad emission bands that can be detected: a dominating red band at 650 nm, a yellow band at about 570 nm, and a blue band of mostly low intensity at about 450 nm. Further, wavelength discrimination (in the form of CL spectroscopy) enables trace elements or defects to be differentiated. Trace element data from agate samples of aluminum, iron, potassium, sodium, calcium, and uranium indicate that agate formed from near-surface weathering solutions. Conversely, high chromium, antimony, zirconium, and zinc indicate the involvement of hydrothermal solutions during the agate formation process. Aluminum is the most frequent trace element in quartz and chalcedony with concentrations up to a few 1000 ppm. High antimony content in the Ojo de Laguna, Mexico agate indicated the involvement of hydrothermal solutions.

Chalcedony/agate deposited as veins in joints and fractures indicate deposition by ascending hydrothermal waters. Lenses and pods indicate alteration of subsurface basalts, andesites, dacites, and tuffs to chalcedony, chlorite, hematite/goethite/ limonite, celadonite, magnetite, and the ever-present zeolites, especially laumontite, analcime, and thompsonite. Vesicle filling and cylindrical vesicles, and joint fillings indicate deposition by ascending hydrothermal waters and deposition from surface colloidal solutions.

Textures and mineral pseudomorphs are the key to recognizing the type of deposition. Surface deposition is recognizable by especially botryoidal, stalactitical, reniform aggregate, globular, microcollomorphic, and membraneous tube. Hydrothermal deposition is represented by feathery, jigsaw, and cox and comb textures and rhythmic banding.

The amount of banding or type of banding or lack of banding or plume and/ or dendrite formation suggest the age of an agate and its fluid origins as Moxon (2002), Lee (2007), and Rios (2023) suggested. For example, the lack of banding and other surficial textures in Montana moss and Indian Black Skin agates suggest purely hydrothermal fluid origins and a very "young agate," while conversely, the multiple, finely banded Lake Superior agate suggests a very "old" agate. The time for the formation of agates is on the order of 100,000 to one million years as shown by the West Texas agate and the Argentinian agate.

Many of the agate provinces have surface environments that are evaporitic, arid, and subka-like with alkaline lakes with weathering and evaporitic minerals such as hematite/goethite/limonite, sulfates, anhydrite, aragonite, gypsum, barite, magnesite, limestone and calcite and fossils present. What the connection to agate genesis is, is not clear. For example, the Argentinian Puma agate is formed with fossiliferous limestone enclosed.

Opal is also a surface formation phenomenon, forming on the surface both from hydrothermal waters, forming the volcanic type opal, and from the leaching and dissolution of siliceous deposits, such as tuffs, ashes, volcaniclastic sediments, arkosic sandstones, and ignimbrites by surface meteoric waters, forming the sedimentary-type opal. The time for the formation of opal is also approximately one million years, with studies by Zielinski (2005), on uraniferous opal in Nevada and by Chauvire et al. (2015) on gem opal in Ethiopia. Australian opal, however, as determined by Rey (2013), formed from a 30–40 Ma leaching event.

Sedimentary agate, geodes, and chert formed from the creation of siliceous oozes from siliceous fossils, such as diatoms, radiolarians, and sponge spicules. Chert formed from volcanic ashes is indicated by the presence of relict volcanic shards in the chert. Sedimentary agate deposits formed in conjunction with chert. The cherts are frequently colored a reddish-pink indicating a period of oxidizing, weathering, and erosional time, forming an unconformity; thus, indicating that agates are useful as unconformity indicators. The sedimentary agates that formed in North America (Kentucky, South Dakota, and Montana), from the Pennsylvanian Borden deltaic deposits to the Permian Minnelusa and Phosphoria Formations, are similar in appearance with red fortification-type banding, consisting of iron or hematite, an oxide mineral, indicative of and a product of weathering processes, that is ubiquitous in all of the sedimentary agate, and additionally suggests an oxidizing subaerial environment that created an unconformity. Due to their strong similarity, this raises the strong possibility that these agates are all from the same transgressive unconformity and implies subaerial surface conditions for sedimentary agate formation; at least partially.

In comparing the various agate provinces, there are strong similarities in the agates between (1) the Montana Moss agate and the Indian Black Skin agate and (2) the Chihuahuan Mexican agate, the West Texas plume agate, and the Chur, Iran agate. In addition, origins of the silica for the Mexican and Iranian agate suggest that the bentonite deposits in both areas as being the source of the silica for the agates in those areas.

References

Andrew, J. (2008). Cenozoic tectonic history of the Northern Sierra Madre Occidental, Huizopa, Sonora-Chihuahua, Mexico. In J. E. Spenser, & S. R. Titley (Eds.), *Ores and orogenesis: Circum-Pacific tectonics, geologic evolution and ore deposits* (Vol. 22, pp. 517–528). Arizona Geological Society Digest.

Bagas, L. (2005). Geology of the Nullagine 1: 100,000 geological series. *Explanatory notes* (p. 33). Department of Industry and Resources, Geological Survey of Western Australia.

Bailey, C. (2019). Neo Acadian poets in the Blue Ridge. In *The William and Mary Blogs*. http//:wmblogs.wm.edu/cmbail

Bauer, M., & translated by Spenser, L. J. (1904). *Precious stones*. Edited reprint (1968) Dover Publications.

Bell, B. R., & Williamson, I. T. (2002). Tertiary Igneous Activity. In N. Trewin (Ed.), *The geology of Scotland* (4th ed.). The Geological Society. https://doi.org/10.1144/GOS4P.14

Bell, B. R. (2004). The great plume debate: Skye excursion. AGU chapman conference. Figure 1. https://www.MantlePlumes.org/Chapman/SkyeFieldtrip/.html

Bockoven, N. T. (1976). Petrology and volcanic stratigraphy of the El Sueco area, Chihuahua, Mexico (Unpublished PhD dissertation, University of Texas, Austin, Texas).

Bodden, T. J., Bornhorst, T. J., Begue, F., & Deering, C. (2022). Sources of hydrothermal fluids, inferred from oxygen and carbon isotope composition of calcite, Keweenaw Peninsula Native Copper District, Michigan, USA. *Minerals, 12*, 474–502. https://doi.org/10.3390/min12040474

Boellstorff, J. D. (1980). Collecting in Nebraska's glacial deposits. *Rocks and Minerals, 1980*, 103–106.

Bornhorst, T. J., & Barron, R. J. (Eds.). (2013). *Proceedings v. 59, part 2-fieldtrip guidebook. 59th annual meeting* (p. 118). Institute on Lake Superior Geology, Houghton, Michigan.

Bornhorst, T. J., & Rose, W. I. (1994). Self-guided geological fieldtrip to the Keweenaw Peninsula, Michigan. In *Proceedings v. 40, part 2* (p. 185). Institute on Lake Superior Geology.

Braddock, W. A. (2007). Geology of the Jewel Cave SW Quadrangle Custer County, South Dakota. U.S.G.S. Bulletin 1063-G.

Breikreuz, C. (2013). Spherulites and lithophysae—200 years of investigations on high temperature crystallization domains in silica rich volcanic rocks. *Bulletin of Volcanology, 75*(4).

Bristow, J. W., & Saggerson, E. P. (1983). A general account of Karoo vulcanicity in Southern Africa. *Geologische Rundschau, 72*(3), 1015–1060.

Brown, G. F., Schmidt, D. L., & Huffman Jr., A. C. (1989). *Geology of the Arabian Peninsula shield area of Western Saudi Arabia*. U.S. geological survey professional paper 560-A (p. 188).

Browne, M. A. E., Smith, R. A., & Aitken, A. M. (2002). Stratigraphical framework for the Devonian (Old Red Sandstone) rocks of Scotland South of a line from Fort William to Aberdeen. In *British geological survey research report* (RR/01/04, p. 67).

Bryan, S. E., Ewart, A., Stephens, C. J., Parianos, J., & Downes, P. J. (2000). The Whitsunday Volcanic Province, Central Queensland, Australia: Lithological and stratigraphic investigations of a silicic-dominated large igneous province. *Journal of Volcanology and Geothermal Research, 99*, 55–78.

Bryan, S. E., Ferrari, L., Reiners, P. W., Allen, C. M., Petrone, C. M., Ramos-Rosique, A., & Campell, I. H. (2008). New insights into crustal contributions to large-volume rhyolite generation in the Mid-Tertiary Sierra Madre Occidental Province, Mexico, revealed by U-Pb geochronology. *Journal of Petrology, 49*(1), 44–77. https://doi.org/10.1093/petrology/egm070

Bryan, S. E., & Ferrari, L. (2013) Large igneous provinces and silicic large igneous provinces: Progress in our understanding over the last 25 years. *GSA Bulletin, 125*(7–8). https://doi.org/10.1130/B30820.1

Bryan, W. H. (1963). The later history of expanded spherulites. *Journal of the Geological Society of Australia, 10*(part 1), 141–149.

Carney, J. N., Aldiss, D. T., & Lock, M. P. (1994). *The geology of Botswana* (p. 113). Geological Survey Department, Bulletin Series, Bulletin 37. Published by the Director, Geological Survey Department, Private Bag 14, Lobatse, Botswana.

Chauvire, B., Rondeau, B., Alexandre, A., Chamard-Bois, S., La, C., & Mazzero, F. (2019). Pedogenic origin of precious opals from Wegel Tena (Ethiopia): Evidence from trace elements and oxygen isotopes. *Applied Geochemistry, 101*, 127–139. https://doi.org/10.1016/j.apgeochem.2018.12.028

Chestnut Jr., D. R. (1992). Stratigraphic and structural framework of the carboniferous rocks of the Central Appalachian Basin in Kentucky. In *Bulletin 3, series XI, 1992, Kentucky geological survey* (p. 30). University of Kentucky.

Chiesa, S., Civetta, L., DeFino, M., La Volpe, L., & Orsi, G. (1989). The Yemen trap series: Genesis and evolution of a continental flood basalt province. *Journal of Volcanology and Geothermal Research, 36*, 337–350.

Clark, R. (2005, September 10–13). Fairburn Agate: Occurrence in the badlands, grasslands, and black hills of Southwestern South Dakota and Northwestern Nebraska. In D. Kile, T. Michalski, & P. Modreski (Eds.), *Symposium on Agate and cryptocrystalline quartz*. Golden Colorado.

Clemons, R. E. (1998). Geology of the Florida mountains Southwestern New Mexico. In *Memoir 43, New Mexico Bureau of mines and mineral resources* (p. 112). New Mexico Institute of Mining and Technology.

Colburn, R. (2010). *The formation of thundereggs (lithophysae)* (p. 272). Kasper Jasper Press.

Cross, B. I. (2008). Classic Agate deposits of Northern Mexico. *The Mineralogical Record, 39*(6).

Darragh, P. J., Gaskin, A. J., & Sanders, J. V. (1976). Opals. *Scientific American, 234*, 84–95.

Darton, N. H. (1889). Record of North American Geology for 1887 to 1889, Inclusive. U.S.G.S. Bulletin 75.

Darton, N. H. (1901). Part IV-hydrography. Preliminary description of the geology and water resources of the southern half of the Black Hills and adjoining regions in South Dakota and Wyoming. In *Twenty-first annual report of the United States geological survey to the secretary of the interior 1899–1900* (pp. 489–598).

Davidson, M.E., 2014. Zircon Geochronology of Volcanic Rocks from the Trans-Pecos Orogenic Belt, Western Texas: Timing the Cessation of Laramide Folding, Uplift, and Post-Slab Ignimbrite Flare-Ups, unpublished M.S. thesis, University of Houston, Houston, Texas, 202 pp.

Dayvault, R. D., & Hatch, S. H. (2005). Cycads: From the Upper Jurassic and Lower Cretaceous Rocks of Southeastern Utah. *Rocks and Minerals, 89*(6), 412–432.

Dejonghe, L., Darras, B., Hughes, G., Muchez, P., Scoates, J., & Weis, D. (2002). Isotopic and Fluid-Inclusion Constraints on the Formation of Polymetallic Vein Deposits in the Central Argentinian Patagonia. *Mineralium Deposita, 37*, 158–172.

DGadmin. (2022, March 15). The Andesite Rose near Freisen. DGGV Digital Geology. https://digitalgeology.de/en/the-andesite-rose

Duex, T. W., & Henry, C. D. (1981). Volcanic geology and mineralization in the Chinati Caldera Complex, Trans-Pecos Texas. In *Geological circular 81–2, Bureau of economic geology* (p. 14). The University of Texas at Austin.

Dumanska-Slowik, M., Powolny, T., Sikorska-Jaworowska, M., Gawel, A., Kogut, L., & Polonski, K. (2018). Characteristics and origins of Agates from Ploczki Gorne (Lower Silesia, Poland): A combined microscopic, micro-Ramen, and cathodoluminescence study. *Spectrochimica Acta Part A: Molecular Spectroscopy, 192*(5), 6–15. https://doi.org/10.1016/j.saa.2017.11.005

Dutkiewicz, A., Landgrebe, T., & Rey, P. (2015). Origins of Silica and fingerprints of Australian sedimentary opals. *Gondwana Research, 27*, 786–795.

Emeleus, C. H., & Bell, B. R. (2005). British regional geology: The Palaeogene volcanic districts of Scotland (4th ed.). British Geological Survey.

Ettensohn, F. R., Lierman, R. T., & Mason, C. E. (2009). Upper Devonian-Lower Mississippian Clastic Rocks in Northeastern Kentucky: Evidence for Acadian Alpine Glaciation and Models for Source-Rock and Reservoir-Rock Development in the Eastern United States (p. 49). American Institute of Professional Geologists-Kentucky Section, Spring Field Trip, April 18, 2009.

Ettensohn, F. R., Lierman, R. T., Udgata, D. B. P., & Mason, C. E. (2012). The Early-Middle Mississippian Borden-Grainger-Ft. Payne Delta/Basin Complex: Field Evidence for Delta Sedimentation, Basin Starvation, Mud-Mound Genesis, and Tectonism During the Neoacadian Orogeny. *The Geological Society of America, Field Guide, 29*, 345–395.

Ewart, A., Marsh, J. S., Milner, S. C., Duncan, A. R., Kamber, B.S., & Armstrong, R. A. (2004). Petrology and geochemistry of early cretaceous bimodal continental flood volcanism of the NW Etendeka, Namibia. Part 1: Introduction, Mafic Lavas and Re-Evaluation of Mantle Source Components. *Journal of Petrology, 45*(1), 59–105. https://doi.org/10.1093/petrology/egg083

Fagnan, B. A. (2009). Geologic Map of the Jewel Cave Quadrangle, South Dakota, 7.5 Minute Series Geologic Quadrangle Map 9. Department of Environment and Natural Resources, South Dakota Geological Survey.

Fallick, A., Jocelyn, J., Donnelly, T., Guy, M., & Behan, C. (1985). Origin of Agates in volcanic rocks from Scotland. *Nature, 313*, 672–674.

Faust, G. T. (1975). A review and interpretation of the geologic setting of the Watchung basalt flows, New Jersey. United States Geological Survey (Professional paper 864-A, p. 42).

Fenner, C. N. (1910). The Watchung basalt and the Paragenesis of its zeolites and other secondary minerals. *Annals N. Y. Academy of Sciences, xx*(2, Part II), 93–187.

Ferrara, V., & Pappalardo, G. (2004). Sea water intrusion in the coastal aquifers of South-Eastern Sicily (Italy). In C. Araguas (Ed.), *18 SWIM conference, Cartagena, Spain* (pp 729–743). IGME.

Frankie, W. T., & Jacobson, R. J. (1998). *Guide to the geology of the Hamilton-Warsaw area, Hancock County, Illinois. Fieldtrip guidebook* (p. 58). Department of Natural Resources.

Ghorbani, M. (2013). Section 3.0 geological situation of Iran. In *The economic geology of Iran* (pp. 45–64).

Gilg, H. A., Morteani, G., Kostitsyn, Y., Preinfalk, C., Gatter, I., & Strieder, A. J. (2003). Genesis of Amethyst Geodes in Basaltic Rocks of the Serra Geral Formation (Ametista do Sul, Rio Grande do Sul, Brazil): A fluid inclusion, REE, oxygen, carbon, and Sr isotope study on basalt, quartz, and calcite. *Mineralium Deposita, 38*, 1009–1025.

Golanka, J., Porebski, S. J., Barmuta, J., Papiernik, B., Bebenek, S., Barmuta, M., Botor, D., Pietsch, K., & Slomka, T. (2019). Paleozoic palaeography of the East European Craton (Poland) in the framework of global plate tectonics. *Annales Societatis Geologorum Poloniae, 89*, 381–403.

Goldich, S. S., & Elms, M. A. (1949). Stratigraphy and petrology of Buck Hill Quadrangle, Texas. *Geological Society of America Bulletin, 60*(7), 1133–1182.

Gooday, B. (2018). Arran's volcanic past-classic geology and new ideas. In A. Stone, & B. McIntosh (Eds.), *The Edinburgh geologist* (No. 64, pp. 13–19).

Gooday, R. J., Brown, D. J., Goodenough, K. M., & Kerr, A. C. (2018). A Proximal record of caldera-forming eruptions: The stratigraphy, eruptive history and collapse of the Palaeogene Arran Caldera, Western Scotland. *Bulletin of Volcanology, 80*, 70–92.

Gordon, C. H. (1895). Stratigraphy of the Saint Louis and Warsaw Formations in Southeastern Iowa. *The Journal of Geology, 3*(3), 289–311.

Götze, J., Nasdala, L., Kleeberg, R., & Wenzl, M. (1998). Occurrence and distribution of "Moganite" in Agate/chalcedony: A combined micro-Raman, Rietveld, and cathodoluminescence study. *Contributions to Mineralogy and Petrology, 133*(1), 96–105.

Götze, J., Mockel, J., Kempe, U., & Kapitonov, I. (2009). Characteristics and origin of Agates in sedimentary rocks from the Dryhead area, Montana, USA. *Mineralogical Magazine, 73*(4), 673–690.

Götze, J., Gaft, M., & Mockel, R. (2015). Uranium and uranyl luminescence in Agate/chalcedony. *Mineralogical Magazine, 79*(4), 983–993.

Götze, J., Moeckel, R., Vennemann, T., & Muller, A. (2016). Origins and geochemistry of Agates in Permian volcanic rocks of the sub-Erzgebirge basin, Saxony (Germany). *Chemical Geology, 428*.

Götze, J., Moeckel, R., & Pan, Y. (2020). Mineralogy, geochemistry and genesis of Agate—a review. *Minerals, 10*, 1037. https://doi.org/10.3390/min.10.10111037. www.mdpi.com/journals/minerals

Götze, J., Moekel, R., Pan, Y., & Muller, A. (2023). Geochemistry and formation of Agate-bearing lithophysae in lower Permian volcanics of the NW-Saxonian basin (Germany). *Mineralogy and Petrology.* https://doi.org/10.1007/s00710-023-00841-2

Graetsch, H., Florke, O. W., & Miehe, G. (1985). The nature of water in chalcedony and opal-C from Brazilian Agate geodes. *Physics and Chemistry of Minerals, 12*, 300–306.

Greb, S., Potter, P., Knox, L., Stapor Jr., F., Meyer, D., & Ausich, W. (2011). Geology of the Fort Payne Formation in South-Central Kentucky and Tennessee. Kentucky Geological Survey. https://www.uky.edu/KGS/geoky/fieldtrip/fort_payne/index.htm

Greb, S. F. (2012). *How geodes and Agates form* (p. 1). University of Kentucky.

Green, J. C. (1979). Fieldtrip guidebook for the Keweenawan (Upper Precambrian) North Shore Volcanic Group, Minnesota. Prepared for the Annual Meeting of the Geological Society of America North-Central Section and the Institute on Lake Superior Geology, Duluth, Minnesota, 1979. Minnesota Geological Survey, University of Minnesota, St. Paul, Minnesota, Guidebook Series No. 11 (p. 22).

Griswold, G. B. (1961). Mineral Deposits of Luna County, New Mexico. Bulletin 72, State Bureau of Mines and Mineral Resources, New Mexico Institute of Mining and Technology, Campus Station, Socorro, New Mexico (p. 157).

Guilbert, J. M., & Park Jr., C. F. (1986). The geology of ore deposits (p. 985). W.H. Freeman and Company.

Habermehl, M. A. (2002). Hydrogeology hydrochemistry and isotope hydrology of the Great Artesian basin. In *Proceedings of the international association of hydrogeologists (IAH)-international groundwater conference-"Balancing the Groundwater Budget", Darwin, 12–17 May 2002*. International Association of Hydrogeologists.

Hansley, P. L., & Sheppard, R. A. (1993). Distribution and properties of clinoptilolite-bearing tuffs in the Upper Jurassic Morrison formation on the Ute Mountain Reservation, Southwestern Colorado and Northwestern New Mexico (p. 11). U.S. Geological Survey Bulletin 2061-A.

Harris, C. (1989). Oxygen-isotope zonation of Agates from Karoo volcanics of the Skeleton Coast, Namibia. *American Mineralogist, 74*, 476–481.

Harris, C. (1990). Oxygen-isotope zonation of Agates from Karoo volcanics of the Skeleton Coast, Namibia: Reply. *American Mineralogist, 75*, 1207–1208.

Harris, C. (1995). The oxygen isotope geochemistry of the Karoo and Etendeka volcanic provinces of Southern Africa. *Suid-Afrikaanse Tydskrif vir Geologie, 98*(2), 126–139.

Hartman, A., & Lloyd, A. (Eds.). (2012). *Nova scotia field guide* (p. 86). Columbia University, Department of Earth and Environmental Sciences. https://eesc.columbia.edu/sites/default/files/content/Field/Guides/NovaScotiaFieldGuide-2.pdf

Hay, R. L. (1963). *Stratigraphy and zeolitic diagenesis of the John Day formation of Oregon.* University of California Publications in Geological Sciences.

Hayes, J. B. (1964). Geodes and concretions from the Mississippian Warsaw formation, Keokuk Region, Iowa, Illinois, Missouri. *Journal of Sedimentary Petrology, 34*(1), 123–133.

Heaney, P. J. (1993). A proposed mechanism for the growth of chalcedony. *Contributions to Mineralogy and Petrology, 115*, 66–74.

Heaney, P. J. (1995). Moganite as an indicator of vanished evaporites: A testament reborn. *Journal of Sedimentary Petrology A, 65*, 633–688.

Heeremans, M., Faleide, J. I., & Larsen, B. T. (2004). Late Carboniferous-Permian of NW Europe: An Introduction to a New Regional Map. In M. Wilson, E.-R. Neumann, G. R. Davies, M. J. Timmerman, M. Heeremans, & Larsen, B. T. (Eds.) *Permo-carboniferous magmatism and rifting in Europe* (Vol. 223, pp. 75–88). Geological Society, London, Special Publications.

Henderson, B., Collins, W., Murphy, J., & Hand, M. (2018). A Hafnium isotopic record of magmatic arcs and continental growth in the Iapetus Ocean: The contrasting evolution of Ganderia and the Peri-Laurentian Margin. *Gondwana Research, 58*. https://doi.org/10.1016/j.gr.2018.02.015

Herman, G. C. (2010). Chapter F. Hydrogeology and Borehole Geophysics of Fractured-Bedrock Aquifers, Newark Basin, New Jersey. In G. C. Herman, & M. E. Serbes (Eds.), *Contributions to the geology and hydrogeology of the Newark basin, New Jersey*. Geological Survey Bulletin 77, State of New Jersey, Department of Environmental Protection, Water Resources Management, New Jersey Geological Survey (p. 45).

Hess, D. F. (1976). Geodes—Occurrence, origin, and mineralogy. In W. A. McCracken (Ed.), *40th annual tri-state field conference* (p. 69). Department of Geology, Western Illinois University.

Hesse, R. (1989). Silica diagenesis: Origin of inorganic and replacement cherts (abs). *Earth Science Reviews, 26*, 253–284.

Hoffmann, U., Breitkreuz, C., Breiter, K., Sergeev, S., Stanek, K., & Tichomirowa, M. (2013). Carboniferous-Permian volcanic evolution in central Europe-U/Pb ages of volcanic rocks in Saxony (Germany) and Northern Bohemia (Czech Republic). *International Journal of Earth Sciences, 102*(1), 73–99. https://doi.org/10.1007/s00531-012-0791-2

Hoisl, T. B. (guide book Ed.), & Green, J. C. (program chair and abstract Ed.). (1989, May). *Part 2 field trip guidebook*. Institute on Lake Superior Geology Proceedings (Vol. 35, Parts A, B, C, and D).

Holzhey, G. (2001). Contribution to petrochemical-mineralogical characterization of alteration processes within the marginal facies of rhyolitic volcanics of lower Permian age, Thuringian forest, Germany. *Chemie der Erde, 61*, 149–186.

Horton, D. (2002). Australian sedimentary opal-why is Australia unique? (p. 11). www.opals.com.au

Howard, P. (1996). Agate Creek Agate. *The Australian Gemmologist, 19*, 215–220.

Huber, N. K. (1975). *The geologic story of Isle Royale National Park* (p. 66). U.S. Geological Survey Bulletin 1309.

Jain, P., Coetzee, S., & Diskin, S. (2013). Optical features, microstructure, and microanalysis of Botswana Agates. *Botswana Notes and Records, 45*, 111–125.

Jochems, A., Hobbs, K., & Mustoe, G. (2022). Fossil wood in the Upper Santa Fe group, South-Central New Mexico: Implications for mineralization style and paragenesis. In *New Mexico geological society guidebook, 72nd fall field conference* (pp. 295–304).

Jones, C. R., & Hepworth, J. V. (1973). Geological map of Botswana. Director of Geological Survey and Mines with the Authority of the Ministry of Commerce, Industry, and Water Affairs Gaborone.

Jourdan, F., Feraud, G., & Bertrand, H. (2006). *Basement control on Dyke distribution in large igneous provinces: Case study of the Karoo Triple Junction*. https://www.mantleplumes.org/Webpage-PDFs/Karoo.pdf

Juchem, P. L. (2014). *Gem materials in Rio Grande do Sul State, Brazil—A field trip guide*. https://api.semanticscholar.org/CorpusID:132581618

Juchem, P. L., de Brum, T. M. M., Fischer, A. C., Liccardo, A., & Chodur, N. L. (2011). *Geology of gemstone deposits in South Brazil*. app.ingcmmet.gob.pe.

Keller, G., Adatte, T., Bajpai, S., Mohabey, D. M., Widdowson, M., Khosla, A., Sharma, R., Khosla, S. C., Gertsch, B., Fleitmann, D., & Sahni, A. (2009). K-T transition in Deccan Traps of central India marks major marine seaway across India. *Earth and Planetary Science Letters, 282*, 10–23.

Keller, P. C. (1977). *Geology of the Sierra del Gallego Area, Chihuahua, Mexico* (Unpublished PhD thesis, University of Texas, Austin, Texas).

Keller, P. C., Bockoven, N. T., & McDowell, F. W. (1982). Tertiary volcanic history of the Sierra del Gallego area, Chihuahua, Mexico. *Geological Society of America Bulletin, 93*, 300–314.

Khalili, M., & Malekmahmoudi, F. (2011). Geochemical variations during alteration of an andesite-basalt to bentonite in Khur, East of Isfahan, Iran. In M. A. T. M. Brockmans (Ed.), *Proceedings of the 10th international congress for applied mineralogy (ICAM) 2012* (p. 360). Springer Publication.

Kieffer, B., Arndt, N., LaPierre, H., Bastien, F., Bosch, D., Pecker, A., Yirgu, G., Ayalew, D., Weis, D., Jerran, D. A., Keller, F., & Meugniot, C. (2004). Flood and shield basalts from Ethiopia: Magmas from the African Superswell. *Journal of Petrology, 45*(4), 793–834.

Knauth, L. P. (1973). Oxygen and hydrogen isotope ratios in cherts and related rocks (Unpublished PhD thesis, California Institute of Technology, Pasadena, California, p. 368).

Knauth, L. P., & Epstein, S. (1976). Hydrogen and oxygen isotope ratios in Nodular and bedded cherts. *Geochimica et Cosmochimica Acta, 40*, 1095–1108.

Knauth, L. P. (1979). A model for the origin of chert in limestone. *Geology, 7*(x), 274–277.

Knauth, L. P. (1994). Petrogenesis of chert. In P. J. Heaney, C. T. Prewitt, & G. V. Gibbs (Eds.), *Silica, Reviews in mineralogy* (Vol. 29). Mineralogical Society of America.

Kontak, D. J. (2006). Geological map of the North Mountain basalt from Cape Split to Brier Island, with comments on its resource potential. In *Mineral resources branch, report of activities 2005* (pp. 39–66). Nova Scotia Department of Natural Resources, Report ME2006-1.

Krans, S. R., Rooney, T. O., Kappelman, J., Yirgu, G., & Ayalew, D. (2018). From initiation to termination: A petrostratigraphic tour of the Ethiopian low-Ti flood basalt province. *Contributions to Mineralogy and Petrology, 173*, 27–59.

Krishnamurthy, P. (2020). The Deccan volcanic province (DVP), India: A review (part 1: Areal extent and distribution, compositional diversity, flow types and sequences, stratigraphic correlations, dyke swarms and sills, petrography and mineralogy). *Journal of Geological Society of India, 96*(1), 3–104. https://doi.org/10.1007/s12594-020-1501-5

Kshirsagar, P. V., Sheth, H. C., Seaman, S. J., Shaikh, B., Mohite, P., Gurav, T., & Chandrasekharani, D. (2012). Spherulites and thundereggs from pitchstones of the Deccan Traps: Geology, petrochemistry, and emplacement environments. *Bulletin of Volcanology, 74*, 559–577.

Landmesser, M. (1988). Structural characteristics of Agates and their genetic significance. *Neues Jahrbuch Mineralogie Abhundlungen, 159*, 223–235.

Lane, A.C., 1911. The Keweenaw Series of Michigan. Michigan Geological and Biological Survey Publication 6, 983 pp.

Lebedev, L. M. (1967). *Metacolloids in endogenic deposits* (p. 298). Plenum Press.

Lee, D. R. (2007). Characterization of silica minerals in a banded Agate: Implications for Agate genesis and growth mechanisms. *Masters Results, 1*–18.

Lee, D. R. (2007). Characterization and the diagenetic transformation of non-and micro-crystalline silica minerals. *Masters Results, 1*–20.

Liesegang, R. E. (1915). *Die Achate. Dresden and Leipzig* (p. 121). Verlag von Theodor Steinkopff.

Livnat, A. (1983). Metamorphism and copper mineralization of the Portage Lake Lava series, Northern Michigan (PhD thesis. University of Michigan, Ann Arbor Michigan, p. 292).

Longhurst, R. A. (1977). Geologic history of Montana Agate. *Lapidary Journal, 31*(6), 1410–1412.

Lorenz, V., & Haneke, J. (2004). *Relationship between diatremes, dykes, sills, laccoliths, intrusive-extrusive domes, lava flows, and tephra deposits with unconsolidated water-saturated sediments in the Late Variscan Intermontane Saar-Nahe Basin, SW Germany* (Vol. 234, pp. 75–124). Geological Society, London, Special Publications. https://doi.org/10.1144/GSL.SP.2004.234.01.07

Lyle, S. A. (revised by Layzell, A. L.). (2009, revised 2020). Glaciers in Kansas (p. 4). Public Information Circular 28, Kansas Geological Survey, University of Kansas, Lawrence, Kansas.

Machedo, F. B., Rocha Jr., E. R. V., Marques, L. S., Nardy, A. J. R., Zezzo, L. V., & Marteleto, N. S. (2018). Geochemistry of the Northern Parana Continental Flood Basalt (PCFB) province: Implications for regional chemostratigraphy. *Brazilian Journal of Geology, 48*(2), 174–199. https://doi.org/10.1590/2317-488920182011820180098

MacPherson, H. G. (1989). *Agates*. National Museum of Scotland and the British Museum (Natural History), co-publishers (p. 72, w/165 color photographs).

Mahaffey, N., & Finkelman, R. B. (2022). The extraordinary variety and complexity of minerals in a single Keokuk Geode from the Lower Warsaw Formation, Hamilton, Illinois, USA. *Minerals, 2022*(12), 914–934.

Mahmoudi, F. M., & Khalili, M. (2015). Origin and formation qualification of Khur Biabanak Agates, Isfahan province. *Journal of Economic Geology, 6*(2), 277–289, serial number 11.

Mahmoudi, F. M., Khalili, M., & Mirlohi, A. (2013). The origins of the bentonite deposits of Tashtab Mountains (Central Iran): Geological, geochemical, and stable isotope evidence. *Journal of Geopersia, 3*(2), 73–86.

Manassero, M., Zalba, P., & Andreis, R. (2000). Petrology of continental pyroclastic and epiclastic sequences in the Chubut Group (Cretaceous): Los Alteres-Las Plumas Area, Chubut, Patagonia Argentina. *Revista Geologica de Chile, 27*.

Manninen, T., Eerola, T., Makitie, H., Vuori, S., Luttinen, A., Senvano, A., & Manhica, V. (2008). The Karoo volcanic rocks and related intrusions in Southern and Central Mozambique. In Y. Pekkalo, T. Lehto, & H. Makitie (Eds.), *GTK consortium geological surveys in Mozambique 2002–2007* (Special paper 48, pp. 211–250). Geological Survey of Finland.

Marshall, R. R. (1961). Devitrification of natural glass. *GSA Bulletin, 72*(10), 1493–1520.

Martz, J. W., Parker, W. G., Skinner, L., Raucci, J. J., Umhoefer, P., & Blakey, R. C. (2012). Geologic map of Petrified Forest, National Park, Arizona. Arizona Geological Survey Contributed Report CM-12-A (p. 18 p.).

Matheson, E. J., & Frank, T. D. (2020). Phosphorites, Glass Ramps, and Carbonate Factories: The Evolution of an Epicontinental Sea and a Late Paleozoic Upwelling System (Phosphoria Rock Complex). *Sedimentology, 67*(6), 3003–3041.

Mattash, M. A. (2006). *Rocks and minerals of Yemen*. Republic of Yemen.

Maxwell, R. A., Dietrich, J. W. (1970). Correlation of Tertiary Rock Units, West Texas. Report of Investigations-No. 70. Bureau of Economic Geology, University of Texas at Austin, Austin, Texas.

Maxwell, R. A., Dietrich, J. W., Wilson, J. A., & McKnight, J. F. (1972). *Geology of the Big Bend Area Field Trip guidebook* (p. 233). Publication 72–59, West Texas Geological Society, Midland, Texas.

McAnulty, W. N. (1955). Geology of Cathedral Mountain Quadrangle, Brewster County, Texas. Report of Investigations No. 25. Bureau of Economic Geology, The University of Texas at Austin, Texas (p. 48). https://doi.org/10.1130/0016-7606(1955)66[531:GOCMQB]2.0.CO;2

McCann, T., Pascal, C., Timmerman, M. J., Krzywiec, P., Lopez-Gomez, J., Wetzel, L., Krawczyk, C., Rieke, H., & Lamarehe, J. (2006). Post-Variscan (end Carboniferous-early Permian) basin evolution in Western and central Europe. *Geological Society, London, Memoirs, 32*, 355–388. https://doi.org/10.1144/GSL.MEM.2006.032.01.22

McDowell, F. W. (2007). Compiler geologic transect across the Northern Sierra Madre Occidental Volcanic Field, Chihuahua, and Sonora, Mexico. Geological Society of America Digital Map and Chart Series 6 (p. 70).

McLemore, V., & Dunbar, N. (2000). Rockhound State Park and Spring Canyon Recreation Area. *New Mexico Geology, 22*(3), 66–71, 86.

Mendum, J. R. (2012). Late Caledonian (Scandian) and Proto-Variscan (Acadian) Orogenic Events in Scotland (p. 33). British Geological Survey, Edinburgh.

Merguerian, C., & Sanders, J. E. (1993). Duke Geological Laboratory, Trips on the Rocks, Guide-book 19, Delaware River Valley Transect of the Newark Basin, New Jersey, Trip 27, 19 June 1993 (p. 64).

Merino, E. (1995). Genesis of Agates in flood basalts: Twisting of chalcedony fibers and trace element geochemistry. *American Journal of Science, 295*(9), 1156–1176.

Miller, J. D. (Ed.). (2010). Field guide to the geology of Precambrian Iron Formations in the Western Lake Superior Region, Minnesota and Michigan. Precambrian Research Center Professional Workshop Series, Geology, Mineralogy, and Genesis of Precambrian Iron Formations, Oct. 10–16, 2010, (p. 13), University of Minnesota, Duluth.

Modreski, P. J. (1995). Origin of chalcedony nodules in rhyolite from the Peloncillo Mountains, New Mexico. Abstracts with programs, 16th Annual New Mexico Mineral Symposium, Nov. 11–12, 1995, Socorro, New Mexico (pp. 6–8). Online ISSN: 2836-7308.

Morrison, G., Mustard, H., Cody, A., Lisitsin, V., Veracruz, J., & Beams, S. (2019). Metallogenic Study of the Georgetown, Forsayth and Gilbertson Regions, North Queensland. Terra Search Party, Ltd and Klondike Exploration Services, Townsville, Queensland, Australia, Report TS 2019/025.

Morteani, G., Kostitsyn, Y., Preinfalk, C., & Gilg, H. A. (2010). The genesis of the Amethyst geodes at Artigas (Uruguay) and the Paleohydrology of the Guarani Aquifer: Structural, geochemical, oxygen, carbon, strontium isotope and fluid inclusion study. *International Journal of Earth Science (Geol Rundsch), 99*, 927–947.

Moxon, T. (2002). Agate: A study of ageing. *European Journal of Mineralogy, 14*, 1109–1118.

Moxon, T., & Reed, S. J. B. (2006). Agate and chalcedony from igneous and sedimentary hosts aged from 13 to 3480 Ma: A Cathodoluminescence study. *Mineralogical Magazine, 70*(5), 485–498.

Moxon, T., & Palyanova, G. (2020). Agate genesis: A continuing enigma. *Minerals, 10*(12), 953–978.

Mrozik, M., Götze, J., Pan, Y., & Mockel, R. (2023). Mineralogy, geochemistry, and genesis of Agates from Chihuahua, Northern Mexico. *Minerals, 13*, 687–718. https://doi.org/10.3390/min13050687

Mustoe, G. (2015). Late Tertiary Petrified Wood from Nevada, USA: Evidence of multiple silification pathways. *Geosciences, 5*, 286–309.

Mustoe, G., & Dillhoff, T. (2022). Mineralogy of Miocene Petrified Wood from Central Washington State, USA. *Minerals, 12*, 31.

Nazari, M. (2004). Agates and geodes from the Khur Area, Central Iran. *The Australian Gemmologist, 22*(1), 21–28.

Nelson, C. E., Jerram, D. A., Single, R. T., & Hobbs, R. W. (2009). Understanding the Facies Architecture of Flood Basalts and Volcanic Rifted Margins and Its Effect on Geophysical Properties. In *Faroe Islands exploration conference: Proceedings of the 2nd conference* (Suppl. 50, pp. 84–103).

Noggeroth, J. (1850). Uber die Achat-Mandlen in den Melaphyren. *Naturwissenschaft, 3*, 147–162.

Oehler, J. W. (1976). Hydrothermal crystallization of silica gel. *Geological Society of America Bulletin, 87*(10), 1143–1152.

Olsen, P. E. (1980). Triassic and Jurassic Formations of the Newark Basin. In W. Manspeizer (Ed.), *Field studies in New Jersey geology and guide to fieldtrips, 52nd Ann. Mtg* (pp. 2–39). New York State Geological Association, Newark College of Arts and Sciences, Newark, Rutgers University.

Olsen, P. E. (1999). Long period Milankovitch cycles from the Late Triassic and Early Jurassic of Eastern North America and their implications for the calibration of the Early Mesozoic time-scale and the long-term behavior of the planets. *Philosophical Transactions of the Royal Society Series A, 357*.

Ostwald, W. (1896). *Lehrbuch der Allgemeinen Chemie* (Vol. 2, Part 1), Leipzig, Germany.

Ottens, B., Gotze, J., Schuster, R., Krenn, K., Hauzenberger, C., Zsolt, B., & Vennemann, T. (2019). Exceptional multi stage mineralization of secondary minerals in cavities of flood basalts from the Deccan volcanic province, India. *Minerals, 9*, 351. https://doi.org/10.3390/min9060351

Overstreet, W. C., Kiilsgaard, T. H., Grolier, M. J., Schmidt, D. L., Domenico, J. A., Donato, M. M., Botinelly, T., & Harms, T. F., (1985). *Contributions to the geochemistry, economic geology, and geochronology of the Yemen Arab Republic* (p. 115). U.S. Geological Survey, Open-File Report 85-755.

Pabian, R., Jackson, B., Tandy, P., & Cromartie, J. (2005). *Agates: Treasures of the earth* (p. 196). The Natural History Museum.

Pabian, R. K., & Zarins, A. (1991). *Banded Agates: Origins and inclusions* (p. 32). University of Nebraska Conservation and Survey Division, Educational Circular 12.

Parrsboro District Chamber of Commerce. (1980). *A Rockhounds' guide to Parrsboro* (p. 25).

Parthasarathy, G., Kunwar, A. C., & Srinivasan, R. (2001). Occurrence of Morganite-rich chalcedony in Deccan Flood Basalt, Killari, Maharashtra, India. *European Journal of Mineralogy, 13*, 127–134.

Peate, D. W., Montovani, M. S. M., & Hawkesworth, C. J. (1988). Geochemical stratigraphy of the Parana Continental Flood Basalts: Borehole evidence. *Revista Brasileira de Geoclenclas, 18*(2), 212–221.

Peate, D. W. (1997). The Parana-Etendeka province in large igneous provinces: Continental, oceanic, and planetary flood volcanism. *Geophysical Monograph, 100*, 217–245.

Peck, D. L. (1964). Geologic reconnaissance of the antelope-Ashwood Area North Central Oregon: With emphasis on the John Day Formation of Late Oligocene and Early Miocene Age. U.S. Geological Survey Bulletin 1161-D.

Pe-Piper, G., & Miller, L. (2003). Zeolite minerals from the North Shore of the Minas Basin, Nova Scotia. *Atlantic Geology, 38*, 11–28. https://doi.org/10.4138/1252

Petranek, J. (2004). Gravitationally Banded ("Uruguay-Type") Agates in basaltic rocks—Where and when? *Czech Geological Survey, Bulletin of Geosciences, 79*(4), 195–204.

Pick, R., Deniel, C., Coulon, C., Yirgu, G., Hofmann, C., & Ayalew, D. (1998). The Northwestern Ethiopian Plateau Flood Basalts: Classification and Spatial Distribution of Magma Types. *Journal of Volcanology and Geothermal Research, 81*, 91–111.

Powolny, T., Dumanska-Slovik, M., Sikorska-Jaworowska, M., & Wojcik-Bania, M. (2019). Agate mineralization in spilitized Permian Volcanics from "Borowno" Quarry (Lower Silesia, Poland)-Microtextural, Mineralogical and Geochemical Constraints. *Ore Geology Reviews, 114*.

Prothero, D. R., & Lavin, L. (1990). *Chert petrography and its potential as an analytical tool in archaeology* (Centennial Special Vol. 4, Ch. 32, pp. 561–564). Geological Society of America.

Puffer, J. H., Husch, J. M., & Benimoff (1989). The Pallisades Sill and Watchung Basalt Flows, Northern New Jersey and Southeastern New York: A Geological Summary and Field Guide. New Jersey Geological Survey, Open-File Report OFR 92-1 (p. 22).

Queensland Government, Australia, Department of Natural Resources and Mines (1995). Northern Queensland Fossicking-Agate Creek Fossicking Area. www.qld.gov.au/recreation/areas-facilities/fossicking

Ramos, V. A., & Folguera, A. (2011). Payenia volcanic province in the Southern Andes: An appraisal of an exceptional quaternary tectonic setting. *Journal of Volcanology and Geothermal Research, 201*, 53–64. https://doi.org/10.1016/j.jvolgeores.2010.09.008

Randive, K., Chaudhary, S., Dandekar, S., Desmukh, K., Peshve, D., Dora, M. L., & Belyatski, B. (2019). Characterization and genesis of the chalcedony occurring within the Deccan Lava Flows of the LIT Hill, Nagpur, India. *Journal of Earth System Science, 128*, 192.

Rey, P. F. (2013). Opalisation of the Great Artesian Basin: An Australian story with a Martian twist. *Australian Journal of Earth Sciences, 60*, 291–314. https://doi.org/10.1080/08120099.2013.784219

Richards, P. (1955). Geology of the Big Horn Canyon-Hardin Area, Montana and Wyoming. United States Geological Survey, Bulletin 1026 (p. 93).

Richter, S., Götze, J., Niemeyer, H., & Mockel, R. (2015). Mineralogical investigations of Agate from Cordon de Lila, Chile. *Andean Geology, 42*(3), 386–396.

Rime, V., Foubert, A., Ruch, J., & Kidane, T. (2023). Tectonostratigraphic evolution and significance of the Afar Depression. *Earth-Science Reviews, 244.* https://doi.org/10.1016/j.arscirev.2023.194519

Rios, F. R., Mizusaki, A. M. P., Michelin, C. R. L., & Roderiques, I. C. (2023). Volcaniclastic and epiclastic diagenesis of sandstones associated with volcano-sedimentary deposits from the Upper Jurassic, Lower Cretaceous, Parana Basin, Southern Brazil. *Journal of South American Earth Sciences, 128.* https://doi.org/10.2139/ssrn.4370537

Robles, R., Zierold, T., Feng, Z., Kretzschmar, R., Merbitz, M., Annaker, V., & Schneider, J. (2011). A snapshot of an Early Permian ecosystem preserved by explosive volcanism: New results from the Chemnitz Petrified Forest, Germany. *Palaios, 27,* 814–834.

Ross, C. S. (1941). Origin and geometric form of chalcedony-filled spherulites from Oregon. *American Mineralogist, 26*(12), 727–732.

Ross, C. S., & Smith, R. L. (1961). *Ash-flow tuffs: Their origin, geological relations, and identification* (p. 81). U. S. Geological Survey Professional Paper 366.

Schmitz, G., & Rooyani, F. (1987). *Lesotho geology, geomorphology, soils* (p. 204). The National University of Lesotho. Morija Printing Works—Lesotho.

Shaub, B. M. (1955). Notes on the origin of some Agates and their bearing on a stylolite seam in Petrified Wood. *American Journal of Science, 253,* 117–120.

Shellnutt, J. G. (2013). The Emeishan large igneous province: A synthesis. *Geoscience Frontiers, 5*(3), 1–26. https://doi.org/10.1016/j.gsf.2013.07.003

Shen, M., Lu, Z., & He, X. (2022). Mineralogical and geochemical characteristics of banded Agates from placer deposits: Implications for Agate genesis. *ACS Omega, 7,* 23858–238634.

Sim, S. (1973). Some thoughts about Agate formations of Scotland. *Lapidary Journal,* 620–623

Singhal, B. B. S. (1997). Hydrogeological characteristics of Deccan Trap formations of India. In *Hard rock hydrosystems (proceedings of Rabat symposium S2, May, 1997)* (pp. 75–80, IAHS Publ. No. 241).

Sinotte, S. R. (1969). The Fabulous Keokuk Geodes volume 1 origin, formation, and development in the Mississippian Lower Warsaw Beds of Southeast Iowa and Adjacent States (p. 292). Stephen R. Sinotte, Publisher.

Smart, J., & Senior, B. R. (1980). Cretaceous basins of Northeastern Australia. In R. A. Henderson & P. J. Stephenson (Eds.), *The geology and geophysics of Northeastern Australia* (pp. 315–328). Geological Society of Australia.

Smedes, W., & Prostka, H. J. (1972). *Stratigraphic framework of the Absaroka volcanic supergroup in the Yellowstone National Park Region* (p. 33). U.S. G.S. Professional Paper 729-C.

Smith, A. E. (1997). Geodes from the Warsaw Formation of Missouri, Iowa and Illinois. *Rocks and Minerals, 72*(6), 420–423. https://doi.org/10.1080/00357529709605074

Smith, R. A. (1984). The lithostratigraphy of the Karoo supergroup in Botswana. *Bulletin Geological Survey of Botswana, 26,* 239.

Smith, R. L. (1960). *Zones and zonal variations in welded ash flows* (pp. 149–159). U.S. Geological Survey Professional Paper 354-F.

Soager, N., Holm, P. M., & Llambias, E. J. (2013). Payenia volcanic province, Southern Mendoza, Argentina: OIB Mantle Upwelling in a Backarc Environment. *Chemical Geology, 349–350,* 36–53.

Spagnuolo, M. G., Orts, D. L., Gimenez, M., & Folguera, A. (2015). Payenia quaternary flood basalts (Southern Mendoza, Argentina): Geophysical constraints on their volume. *Geoscience Frontiers,* 1–8.

Starkova, M., Rapprich, V., & Breitkreuz, C. (2011). Variable eruptive styles in an ancient monogenetic volcanic field: Examples from the Permian Levin volcanic field (Krkonose Piedmont Basin, Bohemian Massif). *Journal of Geosciences, 56,* 163–180.

Strieder, A. J., & Heemann, R. (2006). Structural constraints on Parana basalt volcanism and their implications on Agate geode mineralization (Salto do Jacui, RS, Brazil). *Pesquisas em Geociencias, 33*(1), 37–50. https://doi.org/10.22456/1807-9806.19525

Sukheswala, R. N., Avasia, R. K., & Gangopadhyay, M. (1974). Zeolites and associated secondary minerals in the Deccan Traps of Western India. *Mineralogical Magazine, 39*, 658–671.

Surdam, R. C., & Parker, R. D. (1972). Authigenic alumino-silicate minerals in the Tuffaceous rocks of the Green-River Formation, Wyoming. *Geological Society of America Bulletin, 83*, 689–700.

Taleb, T. E. (1971). Sediment properties and depositional environment of the Minnelusa Formation (Permo-Pennsylvanian), Northern Black Hills, South Dakota and Wyoming (Unpublished M.S. thesis, University of Missouri-Rolla).

Tazhizadeh-Farahmand, F., Afsari, N., & Sodoudi, F. (2014). Crustal thickness of Iran inferred from converted waves. *Pure and Applied Geophysics, 172*. https://www.researchgate.net/publication/264859821_Crustal_Thickness_of_Iran_Inferred_from_Converted_Waves

Teboul, P.-A., Durlet, C., Girard, J.-P., Dubois, L., San Miguel, G., Virgone, A., Gaucher, E. C., & Camoin, G. (2019). Diversity and origin of quartz cements in continental carbonates: Example from the Lower Cretaceous rift deposits of the South Atlantic Margin. *Applied Geochemistry, 100*, 22–41.

Thom Jr., W., Hall, G., Wegemann, C., & Moulton, G. (1935). Geology of Big Horn County and the Crow Indian Reservation Montana. United States Geological Survey, Bulletin 856, (p. 200).

Thompson, G. M., Ali, J. R., Song, X., & Jolley, D. W. (2001). Emeishan basalts, SW China: Reappraisal of the formation's type area stratigraphy and a discussion of its significance as a large province. *Journal of the Geological Society, London, 158*, 593–599.

Thoresen, L., & Overlin, S. (Eds.). (2021). *Seventeenth annual Sinkankas symposium—Agate and chalcedony*. Pala International Inc.

Thorne, A. M., & Trendall, A. F. (2001). Geology of the Fortescue Group, Pilbara Craton, Western Australia. Geological Survey of Western Australia, Bulletin 144 (p. 249).

Tripp, R. B. (1959). The mineralogy of Warsaw Formation Geodes. In *Proceedings of the Iowa Academy of Science*, (Vol. 66, No. 1, Annual Issue, Article 47, pp. 350–356).

Udgata, D., & Ettensohn, F. (2020). Middle Mississippian (Late Osagean, early Visean) Floyds Knob Glauconite Interval, Borden and Ft. Payne Formations, Appalachian and Illinois Basins, Kentucky, U.S.A.: Synergistic Influence of Tectonics, Paleoclimate, and Paleogeography. The Geological Society of America, Special Paper 545. https://doi.org/10.1130/2020.2545(07)

Van Gosen, B., Wilson, A., Hammarstrom, J., & Kulik, D. (1996). Mineral resource assessment of the Custer National Forest in the Pryor Mountains, Carbon County, South-Central Montana. United States Geological Survey Open-File Report 96-256.

Van Tuyl, F. M. (1912). The origin of the geodes of the Keokuk Beds. In *Proceedings of the Iowa Academy of Science* (Vol. 19, Annual Issue, Article 29, pp. 169–172).

Volkert, R. A. (2006). Bedrock Geologic Map of the Paterson Quadrangle, Passaic, Essex, and Bergen Counties, New Jersey. Geologic Map Series GM6-06. Department of Environmental Protection, Land Management, New Jersey Geological Survey.

Volkert, R. A. (2007). Bedrock Geologic Map of the Orange Quadrangle Essex, Passaic, Bergen, and Hudson Counties, New Jersey. Geologic Map Series GMS07-1. Department of Environmental Protection, Land Use Management, New Jersey Geological Survey.

Walton, A. W. (1975). Zeolitic diagenesis in Oligocene volcanic sediments, Trans-Pecos Texas. *Geological Society of America Bulletin, 86*, 615–624.

Walton, A. W., & Henry, C. D. (1979). Cenozoic Geology of the Trans-Pecos Volcanic Field of Texas. Guidebook 19, Bureau of Economic Geology, The University of Texas at Austin, Austin, Texas (p. 193).

Wang, X., Cao, J., Zhang, B., Liao, Z., Zhang, B., Liu, J., & Shi, J. (2022). Genesis of the Wangpo Bed in the Sichuan Basin: Formation by Eruptions of the Emeishan Large Igneous Province. *Palaeogeography, Palaeoclimatology, and Palaeoecology, 594*. https://doi.org/10.1016/j.palaeo.2022.110935

Wang, Y., & Merino, E. (1990). Self-organizational origin of Agates: Banding, fiber twisting, composition, and dynamic crystallization model. *Geochimica et Gosmochimica Acta, 54*(6), 1627–1638.

Wang, Y., & Merino, E. (1995). Origin and fibrosity and banding in Agates from flood basalts. *American Journal of Science, 295*(1), 49–77.

Weir, G. W., Gualtieri, J. L., & Schlanger, S. O. (1966). Borden Formation (Mississippian) in South and Southeast-Central Kentucky. U.S. Geological Survey Bulletin 1224-F (p. 38).

West Texas Geological Society. (1974). 1974 Field Trip Guide, Publication #74-63: Chihuahua & Sinola Mexico Los Mochis-Topolobampo.

Whitbread, K., Ellen, R., Callaghan, E., Gordon, J. E., & Arkley, S. (2014). East Lothian Geodiversity Audit. British Geological Survey Internal Report, OR/14/063. MediaWiki Publishers.

Wikipedia. (2024, January). *Caledonian Orogeny.* https://en.wikipedia.org/wiki/Caledonian_orogeny

Wilson, J. A. (1980). Geochronology of the Trans-Pecos Texas Volcanic Field. New Mexico Geological Society Guidebook, 31st Field Conference, Trans-Pecos Region (pp. 205–211).

Withjack, M. O., Schlische, L. W., Malinconico, M. L., & Olsen, P. E. (2013). Rift-basin development: Lessons from the Triassic-Jurassic Newark Basin of Eastern North America. *Geological Society, London, Special Publications, 369*, 301–321.

Wolff, R. G., & Huber, N. K. (1973). The Copper Harbor Conglomerate (Middle Keweenawan) on Isle Royale, Michigan, and its Regional Implications. U.S. Geological Survey Professional Paper 754-B (p. 15).

Xu, S., Chen, D., & Shi, X. (2021). Unique Metal Sulfide Inclusion in "Bing Piao" Red Agate from Liangshan, China. *Gems and Gemology, 57*(2).

Yazdani, M. (2019). The position of Agate in the geology of the eastern zone in Iran. Research Gate has not been able to resolve any citations for this publication.

Zakhanova, N., & Goldberg, D. (2021). Mechanical properties of Mesozoic Rift Basin formation. *Geomechanics and Geophysics for Geo-energy and Geo-Resources, 7.*

Zamanian, H., Talefagel, E., Hayatalgheybi, M., Samani, B., & Harris, C. (2020). Geochemical fluid inclusion and O-H-S isotope studies of the Milajerd Au-polymetallic prospect, Central Iran: Implications for ore genesis. *Ore Geology Reviews, 120.* https://doi.org/10.1016/j.oregeo rev.2020.10344

Zarins, A. (1977). Origins and Geologic History of Siliceous Metacolloidal Deposits, Cathedral Mountain Quadrangle, Brewster County, West Texas (Unpublished MS thesis, University of Nebraska, p. 49).

Zarins, A., & Pabian, R. K. (1991). Stratigraphic Distribution and Environments of Deposition Relating to the Origin of Banded Agates. Abstracts with Programs. Geological Society of America. North Central Section. Iowa City, Iowa. April 30–May 1, 1992 (Vol. 24, No. 3, p. 58).

Zeh, A., Gerdes, A., Klemd, R., & Barton, J. M., Jr. (2007). Archean to Proterzoic Crustal Evolution in the Central Zone of the Limpopo Belt (South Africa–Botswana): Constraints from Combined U-Pb and Lu-Hf Isotope Analysis of Zircon. *Journal of Petrology, 48*(8), 1605–1639.

Zhang, X., Ji, L., & He, X. (2020). Gemological Characteristics and Origins of the Zhanguohong Agate from Beipiao, Liaoning Province, China: A Combined Microscopic, X-ray Diffraction, and Ramon Spectroscopic Study. *Minerals, 10*(5), 401. https://doi.org/10.3390/min10050401

Zhou, D., Shi, G., Liu, S., & Wu, B. (2021). Mineralogy and Magnetic Behavior of Yellow to Red Xuanhua-Type Agate and its Indication to the Forming Condition. *Minerals, 11*, 877. https://doi.org/10.3390/min11080877

Zielinski, R. A. (2005). Uraniferous opal, Virgin Valley, Nevada: Contains evidence of its age and origins. In D. Kile, T. Michalski, & P. Modreski, (Eds.), *Symposium on Agate and Cryptocrystalline Quartz, Golden, Colorado, Sept. 10–13, 2005.*